Style and mechanics of liquefaction-related deformation, lower Absaroka Volcanic Supergroup (Eocene), Wyoming

Paul L. Decker*
Department of Geological Sciences
University of Wisconsin–Madison
Madison, Wisconsin 53706

---

*Present address: ARCO Alaska, Inc., P.O. Box 100360, Anchorage, Alaska 99510.

THE GEOLOGICAL SOCIETY OF AMERICA · 1888 ·

SPECIAL PAPER

240

1990

Published by The Geological Society of America, Inc.
3300 Penrose Place, P.O. Box 9140, Boulder, Colorado 80301

GSA Books Science Editor Campbell Craddock

Printed in U.S.A.

Library of Congress Cataloging-in-Publication Data
Decker, Paul L., 1962–
Style and mechanics of liquefaction-related deformation, lower Absaroka Volcanic Supergroup (Eocene), Wyoming / by Paul L. Decker.
p. cm. — (Special paper / Geological Society of America ; 240)
Includes bibliographical references.
ISBN 0-8137-2240-3
1. Rock deformation—Absaroka Volcanic Supergroup (Mont. and Wyo.) 2. Geology, Structural—Absaroka Volcanic Supergroup (Mont. and Wyo.) 3. Geology, Stratigraphic—Eocene. 4. Geology—Absaroka Range Region (Mont. and Wyo.) 5. Absaroka Volcanic Supergroup (Mont. and Wyo.) I. Title. II. Series: Special papers (Geological Society of America) ; 240.
QE604.D43 1990
551.8′09787′4—dc20 90-30299
CIP

Cover Photo: Subsurface liquefaction during early middle Eocene time at depths as much as 180 m (approximately 600 ft) or more is interpreted to have remobilized the massive, coarse epiclastic breccias in the lower half of this exposure. In this case, gravitational adjustments to the loss of strength at this level took the form of diapiric upwelling of the liquefied breccias and locally complex faulting in the more coherent, overlying, well-stratified epiclastic sequence. These units of the lower middle Eocene Wapiti Formation are exposed in a Holocene landslide scar roughly 600 ft high on Hardpan Creek, South Fork Shoshone River Valley.

10 9 8 7 6 5 4 3 2

Contents

Acknowledgments v

Abstract 1

Introduction
  Geographic and geologic setting 2
  Review of previous work 2
    Stratigraphic investigations 2
    Structural investigations 3
  This research 3
    Definition of the problem 3
    Methods and procedures 6

An overview of pre–middle Eocene stratigraphy and geologic history of northwestern Wyoming
  Introduction 6
  Precambrian rock units and events 6
  Paleozoic stratigraphy and events 6
  Mesozoic and lowermost Cenozoic stratigraphy and the Laramide orogeny 7

Stratigraphy of the Absaroka Volcanic Supergroup in the east-central Absaroka Range
  Introduction 7
  Volcanic facies relations 10
  Epiclastic units 10
  Pyroclastic units 16
  Lava flows and minor intrusions 17
  Description and correlation of AVS units 18
  Summary 19

Structural style and kinematics of early middle Eocene deformation in the Wapiti and Aycross Formations
  Introduction 19
  Type 1 domains: Tight to isoclinal overturned folds 21

Type 2 domains: Layer-parallel shearing structures 23
Type 3 domains: Gently dipping truncation surfaces 23
Summary of kinematics in domains of Types 1 through 3 27
Type 4 domains: Liquefaction and chaotic dismemberment 27
Discussion of classification scheme 30

Infinite-slope computer modeling of (Type 1) fold development in the Aycross Formation, Timber Creek area
Introduction 30
Method and procedure 31
Geology 31
Description of the computer model 31
Discussion 31
Conclusions 35

Role of liquefaction in (Type 4) deformation of the Wapiti and Aycross Formations
Introduction 36
Liquidization: Distinguishing among processes 38
Mechanical theory of liquefaction and cyclic mobility 39
Structural evidence of liquefaction in middle Eocene deformation 42
Conditions of deformation 53
Liquefaction and related deformation of historic earthquakes 55
Summary 57

Early Cenozoic deformational history and stratigraphic development
Early Eocene 57
Early middle Eocene 57
Relation between liquefaction-related deformation and Heart Mountain faulting 59
Middle to late Eocene 60
Late Eocene or later 61

Conclusions
Summary of stratigraphic development in the AVS 62
Summary of early middle Eocene structural development 63

Appendix I 64
Appendix II 66
Appendix III 67

References Cited 68

Acknowledgments

Field and laboratory study for this project was funded by the American Association of Petroleum Geologists Grants-in-Aid Committee, the Geological Society of America Committee on Research Grants, the Colorado Scientific Society Memorial Fund Committee, ARCO Oil and Gas Co., and the University of Wisconsin—Madison Department of Geology and Geophysics. The reviews and editing of C. Craddock, R. H. Dott, Jr., H. W. Smedes, and H. J. Melosh have greatly improved this report. The computer modeling described here was carried out with guidance from P. Bosscher. S. W. Bailey provided technical assistance in performing x-ray diffraction analyses, as did Professor E. Glover with electron microprobe (EDA) techniques, both of which were helpful in the identification of clay minerals. Dr. R. W. Allmendinger created and permitted the use of *Stereonet v. 3.0,* a Macintosh computer plotting program. Special thanks go to Phyllis C. Ploudré, C. Craddock, and M. Kennedy, all of whom were helpful in the field phase of this research.

Geological Society of America
Special Paper 240
1990

Style and mechanics of liquefaction-related deformation, lower Absaroka Volcanic Supergroup (Eocene), Wyoming

ABSTRACT

Stratigraphically equivalent sections within the Aycross and lower Wapiti Formations (middle Eocene) in the east-central Absaroka Range of northwest Wyoming locally are intensely deformed in a wide variety of structural styles. Deformation zones are primarily restricted to a well-defined, largely volcaniclastic interval within these two units of the lower Absaroka Volcanic Supergroup (AVS). Unlike the Mesozoic and Paleozoic formations that constitute the major displaced masses of the Heart Mountain, Reef Creek, and South Fork detachment faults in the northern Absaroka Range, the pre-volcanic strata (lower Eocene and older units) in the east-central Absaroka Range appear to have acted as basement during deformation. The rootless character of deformation, and the lack of involvement of younger middle to upper Eocene AVS units, prompted previous workers to propose middle Eocene detachment faulting as the principal process of deformation in the east-central Absaroka Range. However, no source area can be identified for the deformed intervals previously considered "allochthonous," nor has any persistent detachment horizon been discovered. In addition, the structures present represent an exceptionally broad range of structural styles whose geometries imply deformation prior to pervasive lithification, and whose attitudes suggest a nearly random kinematic pattern. Furthermore, the recognition of large-scale liquefaction phenomena in close association with most other deformational features throughout the study area suggests that the volcanic-volcaniclastic pile was more profoundly affected by *in situ* disturbances than by extensive detachment faulting in this area. Apparent lateral transport within local domains is generally limited, although well-developed sheath folds imply locally high shear strains ($\gamma \geqslant 10$). Where subhorizontal mass transport is documented, deformation probably began in response to local instabilities generated by liquefaction at depth. Folding may have proceeded in places at creep rates for decades or centuries after initial liquefaction.

Stratigraphic relations within the AVS in the east-central Absaroka Range are complex, but reveal a pattern of south-southeastward migration of intrusive/extrusive activity throughout middle to late Eocene time. The distal taper of the volcanic-volcaniclastic apron is revealed by southeastward thinning of the Wapiti/Aycross interval as a whole, and individual informally recognized volcanic (Twuf unit) and dominantly epiclastic (Twmb unit) members within this interval. I propose that the deformed basal AVS units in the study area may be the relatively distal stratigraphic equivalents of the Cathedral Cliffs Formation—the basal unit of the AVS in its more proximal northerly reaches. This tentative correlation suggests a genetic link between the spatially and stylistically distinct phenomena of liquefaction-related deformation and Heart Mountain faulting, both of which may stem from a common external high-energy stimulus, probably large-magnitude seismic activity.

INTRODUCTION

The Absaroka volcanic field of northwestern Wyoming and southwestern Montana historically have proven to be fertile ground for research in the realms of volcanology, igneous petrology, stratigraphy, paleontology, structural geology, economic geology, paleomagnetism, and other geological fields. Investigations for this study focused primarily on three areas near the eastern margins of the deeply dissected central Absaroka Range, with the initial intent of expanding on reports of locally intense and enigmatic deformation (Hay, 1954, 1956; Bown, 1982a, b) in the Eocene Absaroka Volcanic Supergroup (AVS). Of necessity, the scope of this research soon widened in an effort to advance simultaneously the understanding of the local and regional stratigraphic framework.

The many different varieties of folds, low-angle normal faults, thrust faults, and chaotic assemblages at the heart of this study are, for the most part, confined to an interval of laterally equivalent strata within the Aycross and lower Wapiti Formations (lower middle Eocene). These structures locally involve the underlying Willwood Formation (lower Eocene), but they do not extend upward into younger units. Because of this stratigraphic confinement, and from the style of other post-Laramide structures nearby, these structures were believed by Bown (1982a, b) to belong to a set of allochthonous masses emplaced by gravitationally driven, listric detachment faulting during middle Eocene time. Other hypotheses advanced to explain parts of this deformation include regional gravity-sliding (Barnes, 1985), and regional gravity-spreading (Hauge, 1982, 1985) as part of the "Absaroka Detachment Complex" (Hauge, 1986). This study demonstrates that, whereas horizontal displacements were locally important, there is a significantly more widespread, but previously unrecognized, contribution from in situ disruption by the effects of excess pore fluid pressures. This report: (1) details the wide variety of structural styles present in the lower Wapiti and Aycross Formations of the east-central Absaroka Range; (2) presents models of deformation formulated in accordance with field observations and the mechanical behavior of unlithified materials; and (3) sets forth an informal stratigraphic nomenclature used in the study area to describe the complex lateral and vertical variation within the AVS.

Geographic and geologic setting

Geological mapping at a scale of 1:20,000 (Plates I through III) was carried out in three key areas near the Greybull and South Fork Shoshone Rivers in the east-central Absaroka Range (Fig. 1), supplemented by geological reconnaissance in much of the surrounding area. Rugged relief throughout the study area affords generally excellent exposure of an Eocene stratigraphic sequence in excess of 6,000 ft (1,800 m) thick deposited above a regional pre-Eocene (Laramide) angular unconformity. However, close inspection and sampling of many spectacular exposures is hindered or precluded entirely by precipitous slopes and sheer cliffs.

The study area is bounded to the west by the roadless Washakie Wilderness, to the north by the equally impassable peaks of Wapiti Ridge, and to the south by the group of mountains that includes Francs Peak, the point of highest elevation in the Absaroka Range. Hence, access to the area is from its eastern and northeastern margins via partially paved and periodically maintained dirt roads from Meeteetse and Cody, Wyoming. For the most part, these roads are confined to the broad valleys cut in the Mesozoic and lower Eocene units, which lie stratigraphically and topographically below the middle Eocene volcanic and volcaniclastic units of primary interest to this study.

Review of previous work

Stratigraphic investigations. Working as a team charged with the immense task of unraveling the geologic story of Yellowstone Park and the adjacent northern Absaroka Range, the early geological party led by Arnold Hague developed a simple, yet meaningful, six-part stratigraphic nomenclature for the Eocene rocks of the region (Hague, 1899; Hague and others, 1896, 1899) based on two sets of volcanic breccias and lava flows:

Late Basalt Sheets
Late Basic Breccia
Late Acid Breccia
Early Basalt Sheets
Early Basic Breccia
Early Acid Breccia

This simple scheme served the cause of broad-based geologic study in Yellowstone Park and the northern Absaroka Range quite well until workers began to pursue more detailed investigations in the middle decades of the twentieth century. Complications also arose in attempting to use this nomenclature in the southern Absaroka Range (Love, 1939), spawning the need for different stratigraphic terms.

Geological mapping, stratigraphic studies, and structural investigations by Love (1934, 1939), Pierce (1941, 1957, 1963a, 1963b, 1978), Pierce and Andrews (1941), Rouse (1933, 1935, 1937, 1940), Hay (1954, 1956), Wilson (1963, 1964a, 1975, 1982), Nelson and Pierce (1968), Rohrer (1966a, b), Rohrer and Obradovich (1969), Smedes and Prostka (1972), Nelson and others (1980), and several other workers were conducted in southerly parts of the Absaroka Range, or in smaller areas within the region studied by Hague's USGS team. A natural, but in some ways unfortunate, consequence of these studies was the proliferation of terms referring to stratigraphic subdivisions that seemed appropriate to each of these workers in the context of their own study areas, but which are difficult to correlate over larger parts of the region. Many of these problems are rooted in the fact that thick, laterally equivalent sections separated by diffuse, poorly known facies transitions were identified by different names, and distinguished in mapping by convenient but arbitrary vertical boundaries.

As some workers renewed the effort toward a regional synthesis in the light of these new stratigraphic schemes, the inevitable confusion over correlations and essential stratigraphic and structural relationships came into full bloom. Much of the recent work in the Absaroka Range (Shive and Pruss, 1977; Bown, 1982a, b; Sundell and Eaton, 1982; Eaton, 1982; Torres and Gingerich, 1983; Sundell, 1984; Hauge, 1985; Barnes, 1985; Torres, 1985; this study) has sought to reevaluate some of the stratigraphic and structural interpretations proposed by workers in previous decades, striving for both regionally valid and locally detailed understanding.

One victim of this reinterpretation is the Pitchfork Formation, originally named by Hay (1954) for exposures encircling the Pitchfork Ranch in the Greybull Valley. This name was applied to the area's thick volcanic-volcaniclastic section, which is coarser than the previously defined Aycross Formation to the south (Love, 1939), but finer than the Early Basic Breccia (Hague and others, 1896, 1899) to the north (now known largely as Wapiti Formation, defined by Nelson and Pierce, 1968). Subsequent work (Smedes and Prostka, 1972; Bown, 1982a, b; Sundell and Eaton, 1982; Eaton, 1982; Sundell, 1984; Hauge, 1985; Barnes, 1985b; this study) reveals that the name "Pitchfork Formation" for this transitional sequence is an unnecessary hindrance to direct correlation of equivalent rock units, and it has been largely abandoned. In general, the primary volcanic and coarser epiclastic parts of the defunct Pitchfork Formation are here assigned to the Wapiti Formation, and the finer portions to the Aycross Formation. The contribution of this study to the understanding of the stratigraphic architecture of the AVS is summarized graphically in Plate IV, and is detailed in the separate section "Stratigraphy of the Absaroka Volcanic Supergroup in the east-central Absaroka Range" below.

Structural investigations. Although mentions of structural aberrations in the rocks of the southern and southeastern Absaroka Range are numerous, relatively few studies have been concerned with them in anything more than incidental fashion. Rouse (1937, 1940) was the first to document widely scattered deformation in the southern Absarokas. Hay (1954, 1956) very clearly defined the unusual structural relationships at three of the localities investigated in the present research, but he did so primarily as an interesting outgrowth of his stratigraphic study of the "Pitchfork Formation," and was unable to settle on a satisfactory explanation. (To his credit, it must be mentioned that Hay recognized the significance of fluid behavior in this deformation, discussed here in terms of in situ liquefaction.)

Wilson (1975) and Barnes (1985) studied a system of late middle(?) Eocene (post–Tepee Trail) gravitational detachment structures in the Wood River area, immediately southeast of the area of this study (Fig. 2). Sundell's (1984) description of the upper middle Eocene Castle Rocks Chaos and Bown's (1982a, b) account of the enormous Plio-Pleistocene Enos Creek–Owl Creek debris avalanche both describe catastrophic gravitational transport of large volumes of volcanic-volcaniclastic rock on the southeastern margins of the Absaroka Range.

The Heart Mountain, South Fork, and Reef Creek detachment faults have long been a topic of controversy in the northern and central parts of the Absaroka Range (for example Dake, 1918; Bucher, 1933, 1935, 1947; Pierce, 1941, 1957, 1963a, b, 1973, 1978, 1985, 1987; Pierce and Nelson, 1973; Nelson and Pierce, 1968; Melosh, 1981; Blackstone, 1985; Hauge, 1982, 1985, 1986). Many workers believe that, regardless of the mechanical enigma these structures present in transporting large slabs of Paleozoic, Mesozoic, and earliest Absaroka volcanic rocks across distances of as much as tens of miles, these faults developed prior to deposition of the middle Eocene Wapiti/Aycross strata considered here. However, this view of the age of these faults is not universally held; Hauge (1982, 1985, 1986) believes that Heart Mountain faulting was noncatastrophic and coeval with deformation by gravity-spreading throughout the Absaroka Range, perhaps occurring over millions of years of Eocene time.

Torres and Gingerich (1983) and Torres (1985) claim that in the valley of the North Fork Shoshone River, Heart Mountain fault blocks lie in structural contact with beds of the lower Aycross Formation, although they are the only workers known to the writer who extend the outcrop pattern of Aycross rocks that far north. Pierce (1986) interprets these same exposures of Heart Mountain fault blocks as resting on pre–Aycross/Wapiti units, probably the Cathedral Cliffs Formation, which he defined (Pierce, 1963b) in the northern Absaroka Range as the earliest-deposited rocks of the Absaroka volcanic sequence, and the only volcanic-volcaniclastic unit involved in Heart Mountain faulting. If the interpretation of Torres and Gingerich (1983) and Torres (1985) is correct, or even if their "lower Aycross" exposures are best reinterpreted as lower Wapiti Formation rocks, it would date movement on the Heart Mountain fault as early middle Eocene, perhaps young enough to coincide with much of the deformation in the South Fork Shoshone River and Greybull River Valleys.

This research

Definition of the problem. Two salient points emerge from the writings of previous workers investigating the structural and stratigraphic problems of the east-central Absaroka Range: (1) there is a need for regional investigations of stratigraphy, depositional processes, and correlations within the AVS; and (2) the processes and timing of widespread, unusual structural development in the lower units of the AVS is poorly understood and merits detailed study. The goals of this study, defined in this light, are: (1) to further understanding of the AVS in the east-central Absaroka Range by improving stratigraphic resolution beyond the formation level by mapping and correlating distinctive informal subunits, by using these stratigraphic refinements to define the temporal relationships among post-Laramide deformational events in the region, by establishing the structural styles and the mechanical controls of sedimentology and stratigraphy upon deformational processes, and by modeling the deformation of AVS units within these temporal, geometric, and mechanical con-

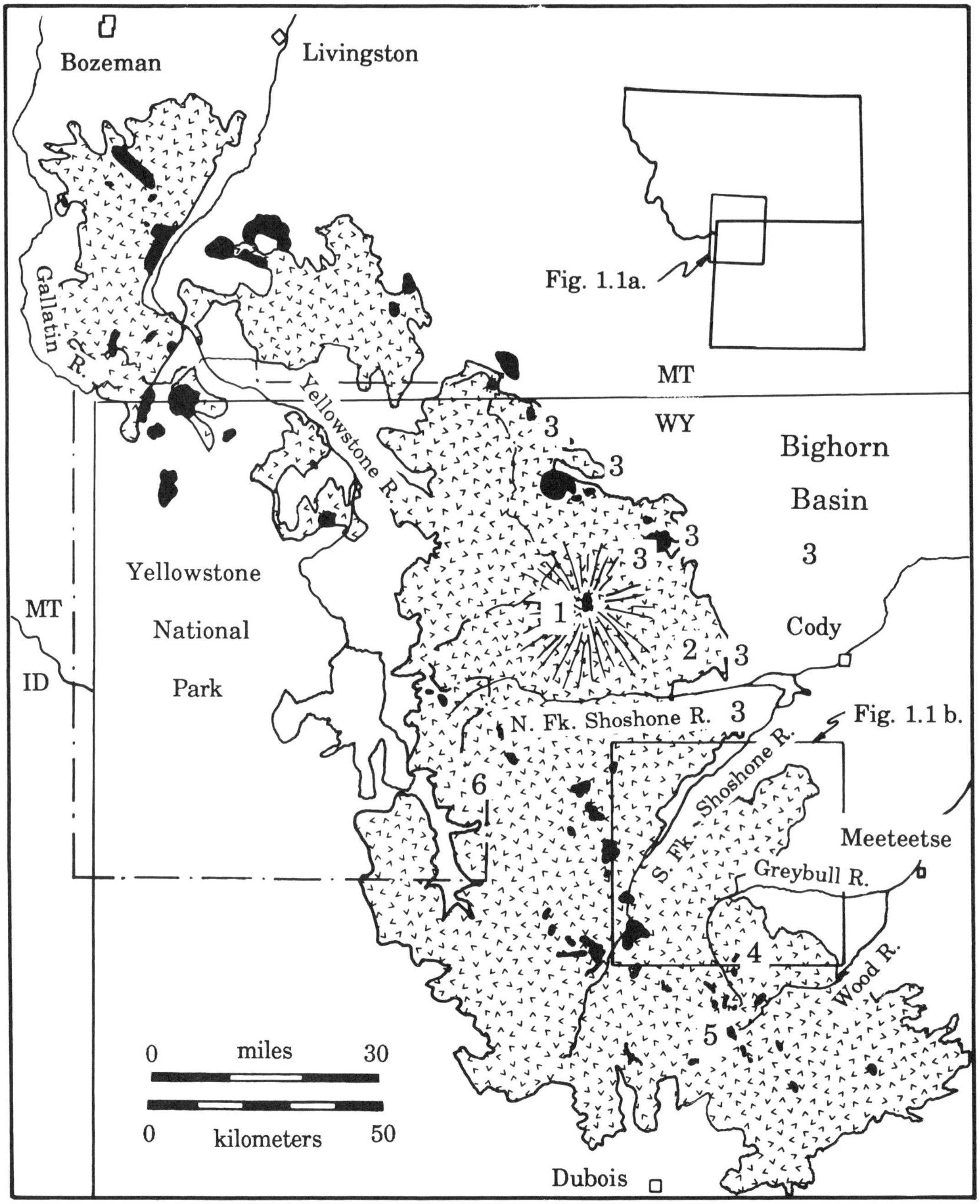

Figure 1a. Regional location map, showing areal extent of the Eocene Absaroka Volcanic Supergroup (AVS) in V-pattern and the principal centers of Absaroka intrusive/extrusive activity in black (redrawn from Chadwick, 1970; Love and Christiansen, 1985). Numbers mark features and locations referred to in text: 1 = Sunlight volcanic enter and accompanying radial dike swarm (simplified); 2 = Jim Mountain; 3 = principal Heart Mountain fault masses, 4 = Francs Peak; 5 = Dollar Mountain; 6 = Eagle Peak. See Figure 1b for map of study area.

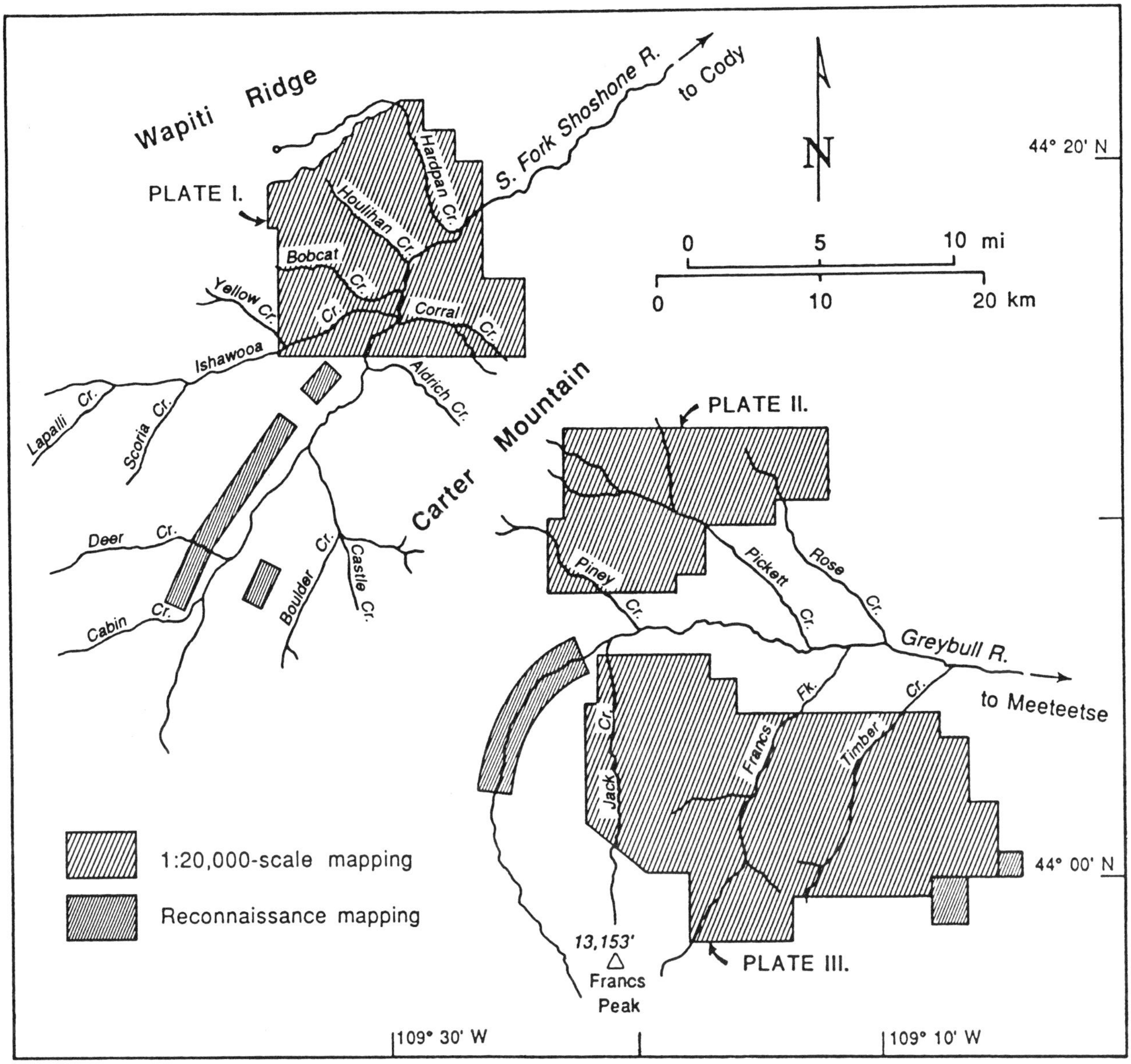

Figure 1b. Map of the study area, showing major geographic features, areas mapped at 1:20,000 (heavy line pattern, see Plates I through III), and areas of geologic reconnaissance mapping (fine line pattern).

Figure 2. View to southeast of Tepee Trail strata, rotated backward over imbricate listric detachment faults between Dick Creek and the Wood River. Because these faults cut Tepee Trail Formation (Thorofare Group), they must be late middle Eocene or younger. This detachment faulting is younger than, and genetically distinct from, the intense deformation confined to Wapiti and Aycross strata (Sunlight Group). Exposures at UTM 12TXD5168 to 5271.

straints; and (2) to advance understanding of structural and stratigraphic processes that may also occur in other volcanic terranes of intermediate composition.

Methods and procedures. The 1:20,000-scale geologic maps of Plates I through III cover three separate areas in the east-central Absaroka Range, with a total combined area of approximately 190 mi^2 (490 km^2). Carried out on USGS topographic base maps photographically enlarged and spliced together at this common scale, mapping was assisted by 1:24,000-scale, color, vertical aerial photographs, and supported by reconnaissance mapping in adjacent areas. Plates I through III correspond to parts of the following USGS quadrangles: Irish Rock, Wapiti, and Sunshine Reservoir (1:62,500-scale 15-minute series), and Dick Creek Lakes, Francs Peak, and Ptarmigan Mountain (1:24,000-scale 7.5-minute series). In the interest of accuracy and efficiency, map locations are referenced in this work using both place names and Universal Transverse Mercator (UTM) grid coordinates. The virtues and workings of the UTM system are explained with an illustrative example in Appendix I.

Laboratory study employed x-ray diffraction (XRD) techniques and electron microprobe energy dispersive analysis (EDA) in addition to optical petrographic microscopy for mineral identification. A broad range of microtextural characteristics was revealed using both optical microscopy and scanning electron microscopy (SEM). Infinite-slope computer modeling helped to clarify the probable process and duration of deformation in a well-developed fold train. For the most part, however, the interpretations and conclusions reported here are derived from well-exposed geologic field relations that received little recent study prior to this investigation.

AN OVERVIEW OF PRE-MIDDLE EOCENE STRATIGRAPHY AND GEOLOGIC HISTORY OF NORTHWESTERN WYOMING

Introduction

The pre-middle Eocene geologic history of northwestern Wyoming is complex and multifaceted, but is obscured in the study area by rocks of the Absaroka Volcanic Supergroup. Whereas Precambrian and pre-Laramide Phanerozoic rock units and events in the local subsurface and in surface exposures of adjoining areas are of major economic and academic interest, they are largely incidental to this study. Hence, these topics are considered only very briefly here, with slightly more emphasis on rock units and events of Late Cretaceous to earliest Tertiary time. Laramide tectonism bears directly on the surface expressions of the few pre-AVS units mapped individually here (Plate III), and the gently dipping subduction of oceanic crust beneath North America during Laramide time is now widely accepted as the ultimate cause of early to mid-Tertiary magmatism throughout the Rocky Mountains, including Absaroka volcanism. The interested reader is referred to previous studies cited here (particularly the reference list of Peterson, 1987) for additional information on the pre-AVS history of northwestern Wyoming and surrounding regions.

Precambrian rock units and events

Well exposed in the core of the Laramide-age Beartooth, Teton, Wind River, Owl Creek, Bighorn, Granite, and Laramie uplifts, Wyoming's Precambrian basement is the nucleus of the Rocky Mountain Archean province. This province is dominated by felsic gneisses and lesser metasedimentary rocks that were metamorphosed at about 2800 Ma and intruded by tonalitic to granodioritic plutons between 2500 and 2760 Ma (Hedge and others, 1986). The oldest known rocks in the province, dated at more than 3300 Ma, occur in the eastern Beartooth Mountains of southwestern Montana (Mueller and others, 1976).

Both compressional and extensional tectonism occurred in this region in Proterozoic time. The Madison mylonite zone (Erslev, 1982, 1983) in the Madison Range-Paradise Valley area (southwestern Montana) records strong crustal convergence at 1500 to 1800 (?) Ma, although mafic dikes with apparent ages of ~2,200 to 1200 Ma are scattered throughout northern and western Wyoming (Hedge and others, 1986), and suggest possible episodic crustal extension during much of Proterozoic time.

Paleozoic stratigraphy and events

The Paleozoic stratigraphy of northwestern Wyoming is dominated by a series of transgressive-regressive epicontinental depositional sequences. Throughout Paleozoic time, this region hovered between the equator and about 10°N, and when submerged, was dominated by shelf carbonate deposition (Dott and

Batten, 1988). Tectonically quiescent conditions prevailed in the Rocky Mountain region from Cambrian through Mississippian time (Craig and others, 1972). The Ancestral Rocky Mountains arose in Colorado and New Mexico during Pennsylvanian time, but this orogenic activity was much subdued in most of Wyoming. The total thickness of Paleozoic rocks on the southern flank of the Beartooth uplift is about 2,800 ft (850 m) (Foose and others, 1986), but reaches as much as 30,000 ft (9,000 m) in western Wyoming and eastern Idaho (Peterson, 1987).

The unconformity-bounded Phanerozoic epicratonic sequences defined by Sloss (1963) constitute a meaningful framework for interpreting the North American geologic record. One of these sequences, the Absaroka Sequence, deserves clarification here. This Pennsylvanian-Triassic sequence has its type succession in the Absaroka Range of northwestern Wyoming, but it is not to be confused with the much younger Absaroka Volcanic Supergroup (Eocene). The Absaroka Sequence's basal unconformity is very complex, largely due to the building of the Ancestral Rocky Mountains (Sloss, 1963) between southern Oklahoma and southeastern Wyoming (Mallory, 1972) during Pennsylvanian time.

Mesozoic and lowermost Cenozoic stratigraphy and the Laramide orogeny

Western Wyoming and southeastern Idaho accumulated a Mesozoic stratigraphic section some 35,000 ft (10,700 m) thick, although this thickness is greatly diminished to the north and east (Peterson, 1987) away from the "Utah-Idaho trough" (Peterson, 1972, p. 180). Mesozoic strata, mostly of shallow-marine and alluvial-plain facies, reflect the variable conditions associated with Cordilleran, Sevier, and Laramide orogenic phases and consequent transgressive-regressive cycles. Bentonite marker beds punctuate the sandstones and shales, which dominated Cretaceous deposition in extensive, shallow, inland seas (McGookey and others, 1972; Foose and others, 1986). Mesozoic paleogeographic reconstructions place northwestern Wyoming at successively higher latitudes, reaching 50 to 55°N by the end of Cretaceous time (Dott and Batten, 1988).

The reader is referred to the excellent maps of Pierce (1970) and Pierce and Nelson (1969) for the outcrop patterns of Mesozoic formations in the study area. These units are grouped in the geological mapping of this study, except in the Greybull River Valley, where the Upper Cretaceous Frontier and Cody Formations define the northwest-trending Four Bear–Willow Draw anticline (Fig. 3; Plate III). This mildly asymmetric, reverse-faulted fold undergoes mild reversals of asymmetry along strike (Plate III), which is not unusual for Laramide structures in the region (Blackstone, 1985).

Upper Cretaceous and Paleocene conglomerates in the southern Yellowstone Park area probably were derived from: (1) highs over ramp anticlines in the Sevier thrust belt (Schmitt, 1987); (2) the hypothetical Targhee uplift, perhaps buried beneath the Snake River Plain (Love, 1956, 1973; Love and Reed, 1971; Love and others, 1978); or (3) the ancestral Teton–Gros Ventre range (Lindsey, 1972). Although some provenance questions remain, a series of angular unconformities clearly demonstrates the synorogenic nature of uppermost Cretaceous and Paleocene clastic units in western Wyoming.

On the east-central flank of the Absaroka Range, the lower Eocene Willwood Formation crosscuts Cretaceous formations at the major Laramide unconformity. The variegated gray and pink siltstones, coarse cross-bedded sandstones, and conglomerates of the Willwood Formation are locally gently folded on the axes of some underlying folds. This late folding is evident on the Four Bear–Willow Draw anticline at Timber Creek in the Greybull River valley (Plate III, ~UTM 12TXD457787 and other nearby areas), and may reflect either a latest stage (mid–early Eocene) pulse of Laramide compression (Van Houten, 1944, p. 206), or tightening of Laramide folds by the intrusion of Absaroka magmas. The latter possibility is suggested by the occurrence of dacite in the core of this anticline, known from drilling in the Four Bear–Willow Draw oil field (Fig. 3a).

Throughout Late Cretaceous to early Eocene time, the mosaic of Sevier highlands and sporadically active Laramide uplifts continued to shed new and reworked clastic sediment (Steidtmann, 1971; Seeland, 1978; Kraus, 1985) into fluvial, lacustrine, and fan delta environments (Lindsey, 1972). Volcanic influences on middle Eocene geography would soon complicate the regional stratigraphy still further in a pattern of local physiographic control on sedimentation, which has continued without reversal until the present.

STRATIGRAPHY OF THE ABSAROKA VOLCANIC SUPERGROUP IN THE EAST-CENTRAL ABSAROKA RANGE

Introduction

Like any terrestrial volcanic terrane, the Absaroka volcanic field in northwestern Wyoming and southern Montana is highly complex—the product of interaction between Earth's internal and surficial processes, between constructive and destructive agents. The stratigraphic framework of the Eocene Absaroka Volcanic Supergroup (AVS) (defined by Smedes and Prostka, 1972) exhibits its horizontal and vertical complexity at many scales. Nonetheless, regional trends are recognized within the well-exposed, 5,000-ft-thick (1,500-m) volcanic-volcaniclastic AVS sequence that provide a meaningful context for smaller local features. The first-order stratigraphic pattern was created by southeastward migration of igneous activity along two subparallel chains of intrusive-extrusive centers (Smedes and Prostka, 1972).

Three groups make up the Absaroka Volcanic Supergroup (Fig. 4a): the Washburn Group (largely middle Eocene, but believed by Smedes and Prostka to be in part lower Eocene), the Sunlight Group (middle Eocene), and the Thorofare Group (middle and upper? Eocene) (Smedes and Prostka, 1972). According to correlations advanced by Smedes and Prostka and subsequent workers, only the upper two groups are represented in

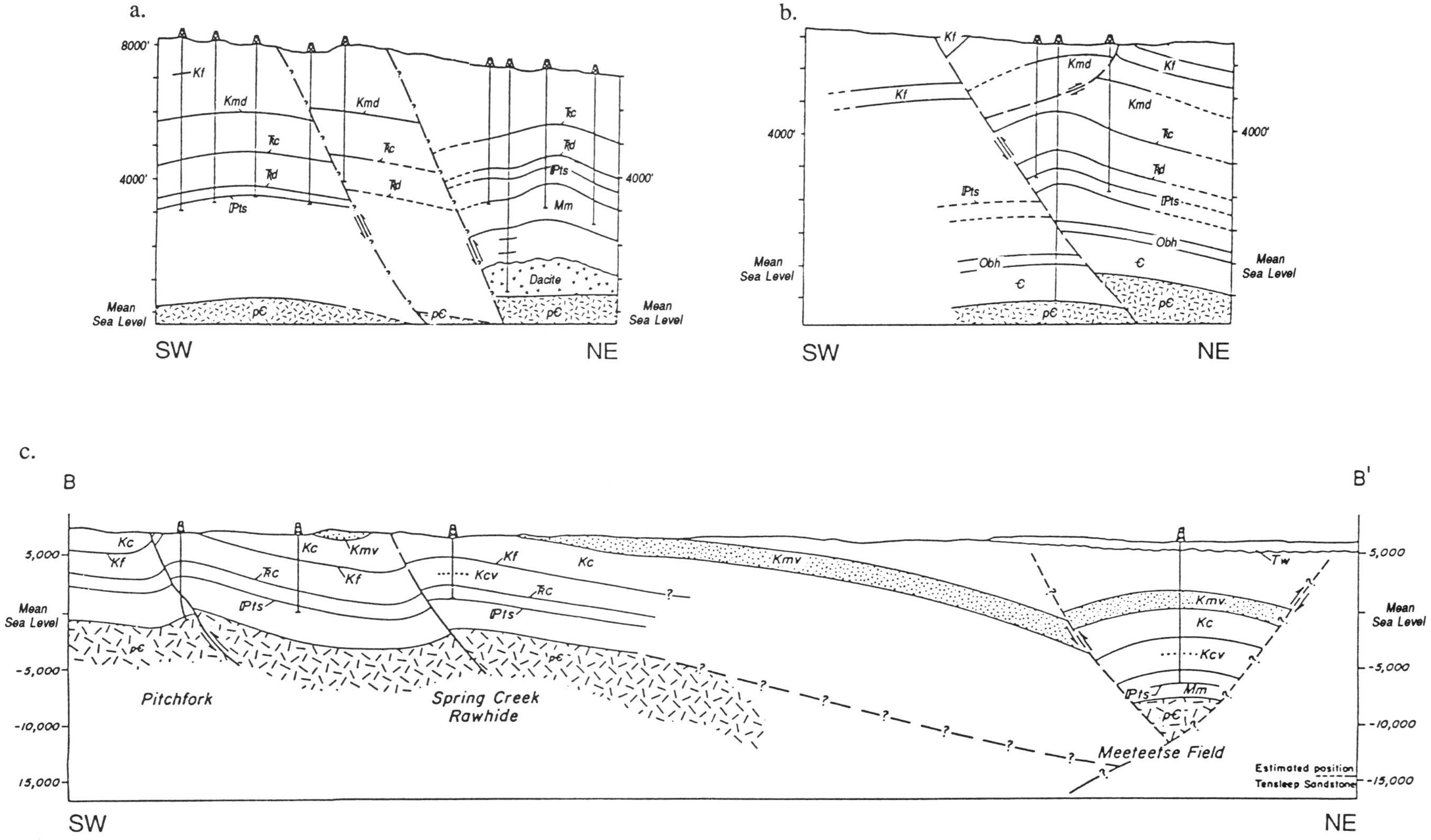

Figure 3. Cross sections showing structural style of Laramide anticlines in western Big Horn Basin, near the study area. Note the controlling influence of reverse faults. (a) Four Bear–Willow Creek anticline. (b) North Sunshine anticline. (c) Meeteetse Field to Pitchfork Field. Simplified from Blackstone (1986).

a.

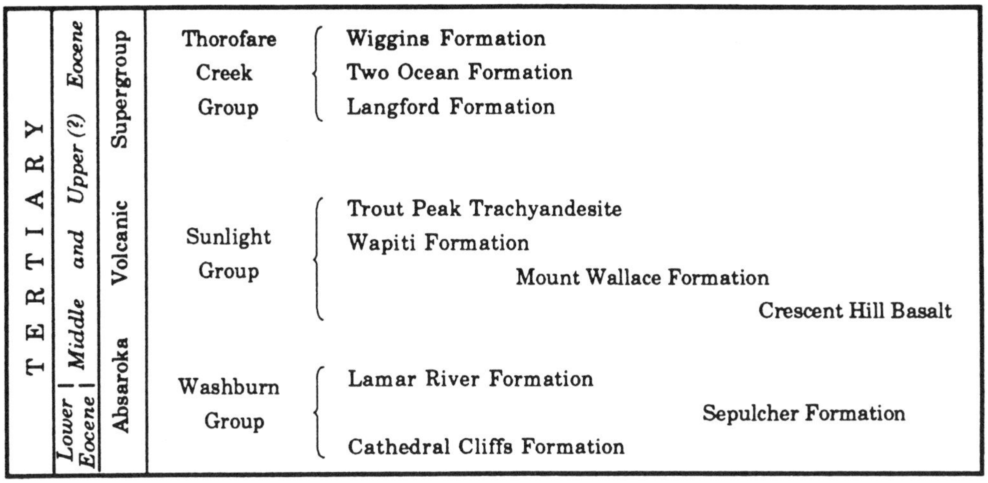

b.

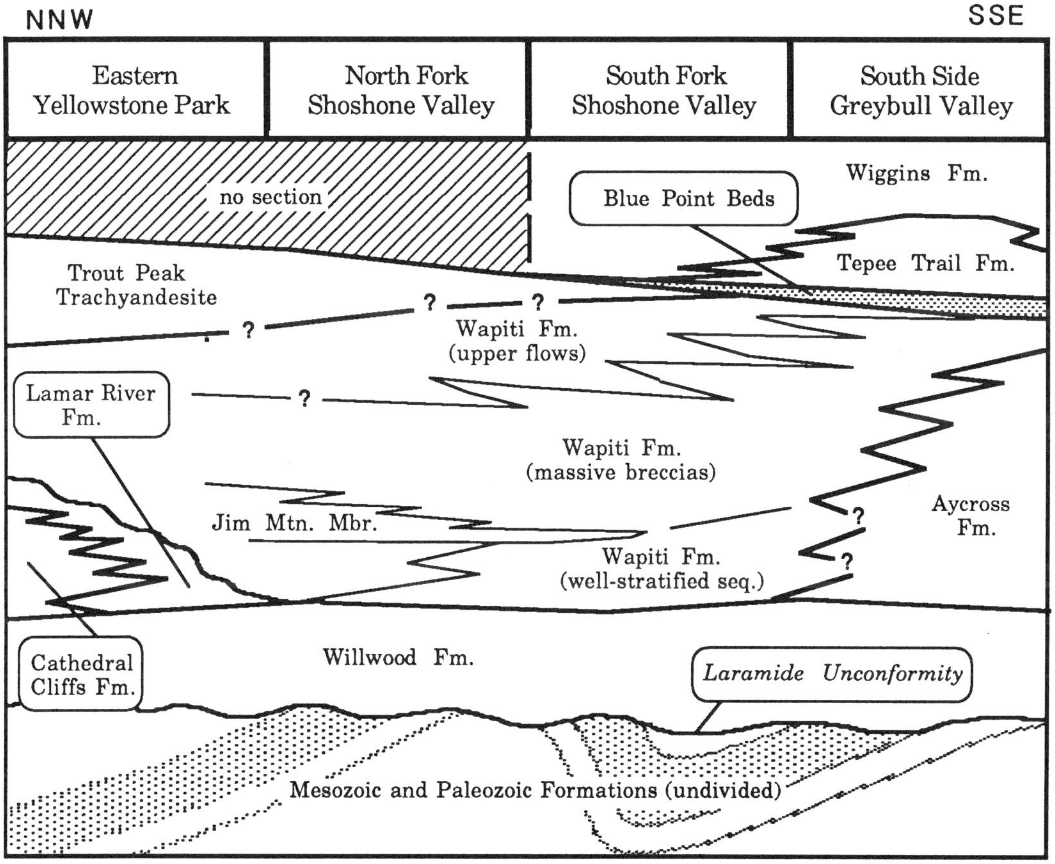

Figure 4. Stratigraphy of the AVS. (a) Units of the AVS and their ages, as originally designated by Smedes and Prostka (1972). The late early Eocene age ascribed to the Cathedral Cliffs Formation is questionable; this unit may be equivalent to lowermost Wapiti/Aycross strata (early middle Eocene age). (b) Schematic relationships among the principal units recognized along the east-central flank of the Absaroka Range from eastern Yellowstone Park to the south side of the Greybull River valley. Lamar River Formation dies out to the east of this section, where Cathedral Cliffs and Wapiti strata may be continuous. See transect of Plate IV for more accurate and detailed relations in South Fork Shoshone and Greybull valley areas.

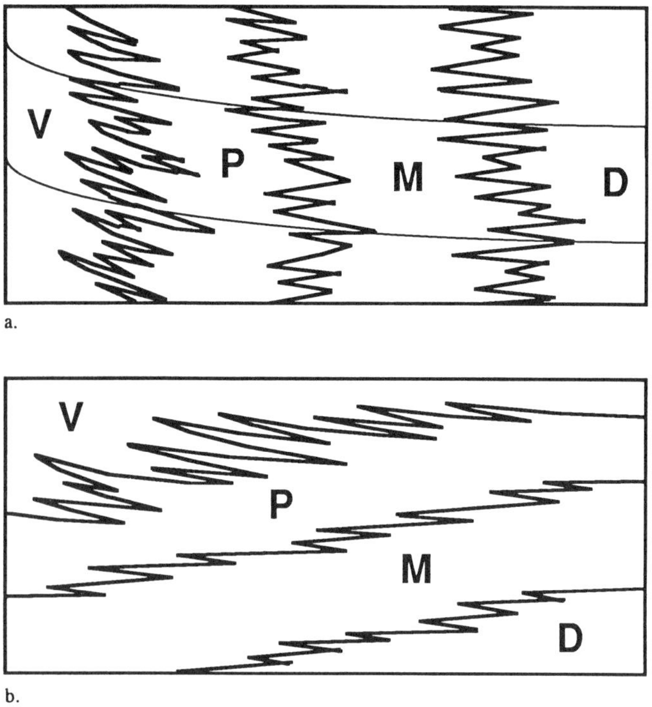

Figure 5. Idealized proximal-distal facies relationships in a volcanic-volcaniclastic pile accumulating around (a) stationary vent complex, and (b) rightward-migrating vent complex. Facies abbreviations: V = vent, p = proximal, m = medial, d = distal. Compare pattern of (b) with Wapiti/Aycross relations in Figure 4b.

the study area. Here, the Bridgerian-age Sunlight Group (Smedes and Prostka, 1972; Sundell and Eaton, 1982) includes, from bottom to top, the laterally equivalent Wapiti and Aycross Formations and the Trout Peak Trachyandesite. The Bridgerian-Uintan-age Thorofare Group (Smedes and Prostka, 1972) includes the Tepee Trail and Wiggins Formations. The Wiggins overlies the Tepee Trail, and in its lower part, interfingers with it (Sundell and Eaton, 1982).

Where possible, Wapiti and Aycross rocks have been informally subdivided here into mappable lithostratigraphic units. Correlations among AVS units in the east-central Absaroka Range are portrayed schematically in Figure 4b, and in greater accuracy and detail in Plate IV. These relations were determined largely in the course of mapping (Plates I through III).

Volcanic facies relations

Stratigraphic successions in volcanic provinces commonly are divided into four facies that reflect distance from the source of volcanic material—the vent, proximal, medial, and distal facies (Fig. 5a). Simpler schemes generalize these environments to just two—the vent and alluvial facies (Parsons, 1974, 1978). During the earliest phase of Absaroka extrusive activity, the vent facies was restricted to the northern Absaroka Range, and the entire study area represented the medial and distal facies. For much of middle Eocene time, the major volcanic source area was the Sunlight Peak vent complex, now revealed as the deeply dissected hub at the center of a classic radial dike swarm. Throughout Wapiti/Aycross time, the study area hosted progressively more proximal facies in sequence, until only its southernmost margin was left untouched by lava flows. The southern limit of flows (Twuf unit of the Wapiti Formation) is traceable from west to east on the geologic map of Plate III in the Jack Creek, Francs Fork, and Timber Creek drainages.

Rose Butte, an informally named landform on the north side of the Greybull River valley near the southeast edge of Plate II (UTM 12TXD4094), has been interpreted as the originally subsurface plug of a small mafic vent (Rouse, 1940; Hay, 1954) active during late Wapiti time, although subsequent detailed mapping reveals that local structural relationships are less symmetrical than originally believed (Fig. 6). The vent hypothesis is supported by the unusually low position of the isolated mafic flow breccias at its summit (Fig. 6; Plate VIII.A). If the butte is a vent, it would constitute the southernmost recognized source of extrusive material during Wapiti Formation time. However, it is possible that the Rose Butte autobreccias are simply an erosional remnant of a valley-filling, Twf/Twfb unit, autobrecciated lava flow that was locally folded along with underlying units during the deformation manifest throughout the study area. In either case, the structural relations at Rose Butte are not attributable solely to vent processes; liquefaction of epiclastic strata there exerts significant, perhaps singular, control on local deformation.

The earliest AVS volcanism recorded in the east-central Absaroka Range was a sequence of explosive eruptions, which eventually overwhelmed the preexisting fluviolacustrine Willwood landscape with felsic pyroclastic debris. Similar pyroclastic strata buried Cambrian through Paleocene units in the northern Teton Range (Love and others, 1978). As volcanism continued, new pyroclastic units and lava flows mingled with the reworked products of earlier eruptions, flanking the eruptive centers with an apron of interfingering lavas and fining-outward mixed volcaniclastic materials. The early middle Eocene saw rapid vertical growth of the AVS coupled with the propagation of eruptive activity to the south-southeast. This development is recorded in the overlapping relation between younger lava and breccia sequences of the proximal volcanic facies to the north-northwest superposed on older, increasingly more distal facies to the south-southeast (Figs. 4b and 5b).

Epiclastic units

Although the term "epiclastic" pertains in its broadest sense to all true sedimentary clastic processes, it is restricted in the present context to clastic materials derived from volcanic rock. These epiclastically reworked volcanic deposits are contrasted here with primary pyroclastic rocks; together, these two terms cover the wide range of rocks that the term "volcaniclastic" describes. Although the AVS as a whole bears the chemical

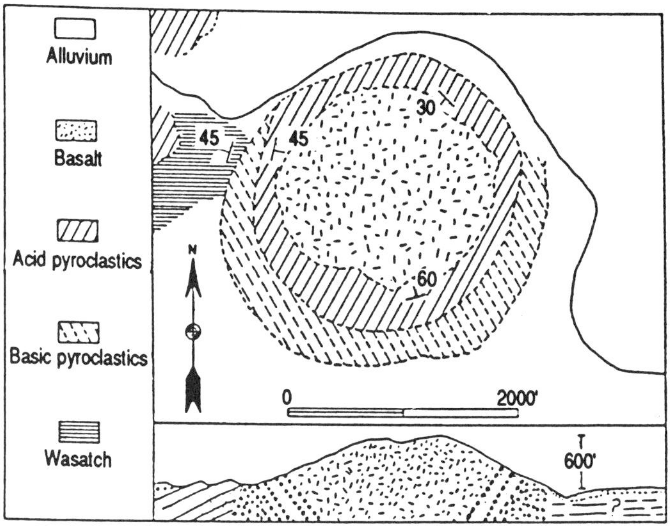

a.

Figure 6. Geologic maps of Rose Butte (~UTM 12TXD4194). (a) Rouse (1940) and (b) Hay (1954) interpreted this butte as a volcanic plug. (c) Mapping for this study suggests an alternative structural interpretation involving an asymmetric, north-plunging anticline-syncline pair. Twtb unit liquefaction breccias (stippled) crosscut the Twca unit andesitic conglomerates, underlying them along the line of section, but overlying them on the knob 0.25 mi south of summit. See also Plates II and VIII.A.

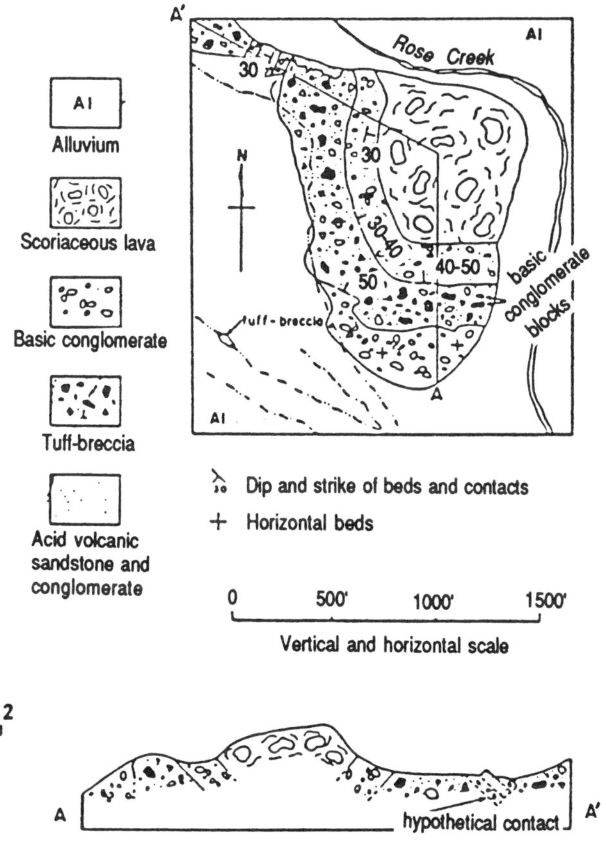

b.

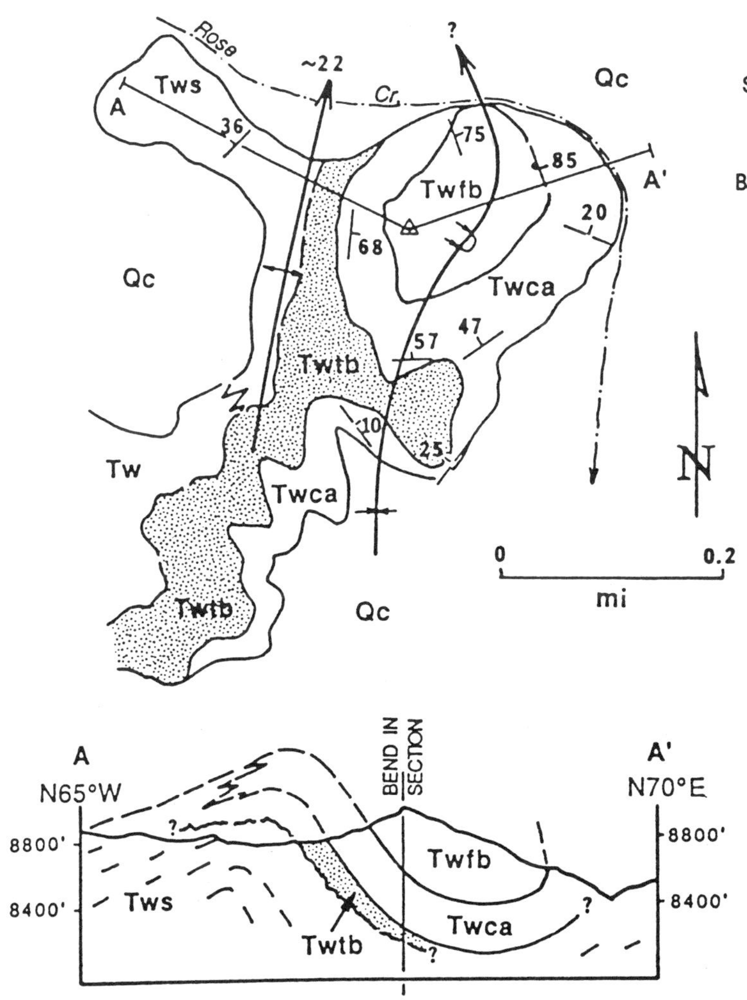

c.

signature of calc-alkaline to alkaline terrestrial volcanism, the more distal volcanic environments embrace few primary volcanic rocks—containing instead mostly immature clastic sedimentary rocks of volcanic provenance.

The Aycross, Tepee Trail, and lower Wapiti Formations are predominantly epiclastic, with lesser volumes of true volcanic rock. On the other hand, epiclastic sediments are largely subordinate or altogether lacking in the upper Wapiti, Trout Peak, and Wiggins Formations, and locally in the Blue Point beds. Epiclastic facies include "normal," well-stratified fluvial and lacustrine deposits as well as large volumes of very poorly sorted, irregularly stratified rocks deposited by lahars and other varieties of "sediment-water flows" (Pierson and Costa, 1987).

Well-stratified epiclastic units. The most abundant epiclastic rock types of the AVS in the study area are volcanic wacke and felsic tuffaceous sandstone, but textures and compositions vary in a wide range that also includes mudstone, siltstone, arkose, conglomeratic sandstone, felsic tuffaceous and andesitic conglomerates, and a broad spectrum of sedimentary breccias. Most of these tend to form well-stratified sequences that lend themselves to interpretation of depositional environments (Bown, 1982a; Eaton, 1982; Sundell and Eaton, 1982; Fritz and Harrison, 1985; Torres, 1985; Barnes, 1985).

In general, well-bedded coarse sand and gravel facies are ascribed to braided and meandering channels, alluvial fans, and overbank environments. Bed geometries are commonly complex, and particle size is highly variable (Fig. 7). A few of these medium- to coarse-grained clastic deposits, especially conglomerates and arkosic sandstones of the Blue Point beds (Tbp unit), are recognizable by their peculiar bright blue-green color. X-ray diffraction analysis of one very pure specimen from the matrix of the lower Tbp unit conglomerate (specimen UW1837-1) reveals that, at least locally, celadonite mica is the chief source of this distinctive color, rather than secondary chlorites (McGrew, 1965, p. 36–39) or chlorites and zeolites combined (Barnes, 1985, p. 67). Electron microprobe (energy dispersive spectroscopy) analyses of three specimens from other units (specimens UW1837-2, 1837-3, and 1837-4; see Appendix III) suggest that celadonite and mixtures of celadonite, chlorite, illite, and montmorillinite contribute to the blue-green and olive green colors of epiclastic matrix materials and mudstone-claystone rocks widely distributed throughout the study area.

In marked contrast to the generally high-energy deposits described above, floodplain, lacustrine and paludal deposits are characterized by dense floral assemblages, rare thin seams of low-grade coal, and a preponderance of fine clastic detritus and bentonite beds. Here, the carbonized impressions of broad palmate leaves (specimen UW1837-5), delicate ferns, reeds (specimen UW1837-6), and scattered lenses of fissile gastropod-bearing marl (specimen UW1837-7) testify to the tropical conditions of middle Eocene time. Although no vertebrate fossils were recovered in this study, the locally abundant Eocene fauna of northwestern Wyoming includes monkeys, crocodiles, turtles,

Figure 7. Complexly channeled alluvial (medial?) facies, Wiggins Formation. Photo shows a roughly 75-ft-thick (25 m) section of interbedded gravels (G) and slightly reworked(?) ash-lapilli tuffs (T). View to WNW on Carter Mountain, just NW of Piney Pass, between headwaters of Piney and Pickett Creeks at UTM 12TXD291928.

and many other mammalian and reptilian species (Bown, 1982a; Eaton, 1982, Torres, 1985).

Field evidence of the volcanic pile's rapid growth is pervasive and well-exposed in these epiclastic units. For example, high on the cliffs in the upper tributaries of Francs Fork (~UTM 12TXD4273), crossbed foresets estimated at more than 80 ft tall document the rapid aggradation of coarse Tepee Trail conglomerates and breccias. The textural immaturity of Tepee Trail epiclastic beds (Fig. 8) further suggests high sedimentation rates. Episodic influxes of pyroclastic debris punctuated the epiclastic record in all depositional environments. In one spectacular example from the upper Blue Point beds on Carter Mountain (UTM 12TXD391945), a very competent stream carrying well-rounded andesitic cobbles and boulders as large as 6 ft (2 m) in diameter was choked by an onslaught of pumice lapilli associated with the Blue Point marker tuff (Fig. 9).

Most individual epiclastic strata and stratigraphic sections in the Wapiti and Aycross Formations exhibit great lateral variability, giving way after only a few miles to pronounced textural and/or compositional change, or tapering out completely. Stratification in the Tepee Trail Formation is unusually uniform by comparison (Fig. 10). Such a lack of horizontal continuity typifies volcaniclastic successions in general. Less common in this environment is the development of a vertically repetitive stratigraphic succession, as in the lower Wapiti Formation at the mouth of Deer Creek Canyon. Here, four nearly identical sequences, each roughly 30 ft thick and containing three distinct beds, are consecutively stacked, implying that local depositional processes, sediment supply, or both changed in consistent cycles.

Massive epiclastic breccias. Thick sections of very poorly bedded to massive breccias occur throughout the proximal facies

Figure 8. Poorly sorted mixtures of angular clasts and finer matrix materials compose a range of immature epiclastic textures in the well-stratified Tepee Trail Formation. View here is of a bedding surface in a tilted erosional lag block composed of laharic breccia interbedded with sandstone. Locality is on ridge crest north of Deer Mountain summit between Francs Fork and Timber Creek at UTM 12TXD525758.

Figure 9. Andesite boulder-lapilli tuff conglomerate, Blue Point beds. Field notebook for scale in left center measures 4.75 × 7.5 in (12 × 19 cm). Grayish-weathering light-colored beds are lapilli tuff; well-rounded cobbles and boulders of andesite occur in lenses of various widths and thicknesses. Even the largest clasts at this exposure are well rounded and measure nearly 6 ft (2 m) in diameter (not shown). View to north on Carter Mountain at UTM 12TXD391945.

of the AVS, as in the middle Wapiti Formation in the South Fork Shoshone Valley (Fig. 11), and these breccias are also prominent as discontinuous lenses in the medial facies (Fig. 12). Lahars (volcanic mudflows) have been identified as the means of their deposition (Rouse, 1937; Hay, 1954; Parsons, 1969, 1978). Because most of these massive breccias are composed of transported sediments derived from preexisting volcanic rock, they are properly classified as epiclastic, although massive pyroclastic breccias, in part vulcanian agglomerates, are also locally significant.

Recognition. Fisher (1960a) considered various criteria for distinguishing epiclastic laharic breccias from vulcanian pyroclastic breccias, and concluded that the two criteria most diagnostic of an epiclastic origin are: (1) the presence of rock fragments or phenocrysts of the same composition in several different stages of alteration, suggesting dissimilar weathering histories prior to deposition; and (2) rounding of sand-size grains caused by abrasion during flow. A second potential problem arises in distinguishing between epiclastic mudflow breccias and autobrecciated lava flows. However, a solution rests in the observation that epiclastic breccias are most likely to be composed of heterogeneous fragments, and may show inverse grading near their base, whereas autobreccias should possess neither of these traits (Parsons, 1969; Fisher and Schminke, 1984).

In general, lahar sediment is heterogeneous, poorly sorted volcanic colluvium or alluvium, mobilized during either hot (eruption-related) or cold floods or mass movement events (Smith, 1986). However, an important distinction must be drawn between "lahar" in general, and "laharic debris flow" in particu-

Figure 10. Stratigraphic relations within the AVS, Timber Creek–Deer Mountain area. Chaotic, massive pyroclastic and epiclastic rocks of the Aycross Formation form partially wooded slopes at base of photo. Bench at left (arrow) is the lower conglomerate of the Blue Point beds (Tbp unit), overlain by white bentonitic slopes of the Blue Point marker tuff to right, and well-stratified coarse epiclastic beds of the upper Tbp unit and Tepee Trail Formation (Tt unit). View to WNW across Timber Creek shows about 1,600 ft (500 m) of relief.

Figure 11. Poorly bedded to massive epiclastic breccias of the proximal-facies, Twmb unit of the Wapiti Formation. Castellated cliff exposures are typical of massive breccias in this area. View to north shows approximately 300 ft (90 m) of relief on cliffs near crest of Wapiti Ridge, upper Hardpan Creek at UTM 12TXE202145.

Figure 12. Massive epiclastic laharic breccias of the medial facies, Wapiti Formation. This exposure consists of two debris flow units, the lower of which is ⩾20 ft (6 m) thick. Note near absence of textural grading in both units. View to northwest on north side of Greybull valley at ~UTM 12TXD3992.

lar. The former term has been variously used since its introduction to the literature (Van Bemmelen, 1949; see Smith, 1986, p. 3); the latter, more specific term is best reserved for volcanic mass movements with the rheologic characteristics detailed below.

A clear understanding of debris-flow mechanics is valuable to this discussion, because other workers have discussed some of the deformation of AVS units considered in this research in terms of closely related mass-wasting processes, such as debris avalanching (Bown, 1982a), landslide-debris flow (Sundell, 1984), catastrophic detachment faulting (Bown, 1982b), and gravity-sliding (Barnes, 1985). In my opinion, debris flow in volcaniclastic sequences is a normal *depositional* process, not to be confused with processes of post-depositional *deformation*.

Rheologic classification of sediment-water flows. It is clear that the term "debris flow," like "lahar," has taken on a variety of meanings in the literature, and is often confused with other types of sediment-water flows. Most current thinking appears to be roughly in accord with that of Smith (1986), who characterized debris flow as en masse, laminar or partly turbulent, water-saturated flow in which clasts are suspended within the matrix. Departure from this view is typically manifested as debate over the particulars of one or another of these defining traits, but some workers add the restriction of high velocity to their definition of debris flow (Takahashi, 1978; Pierson, 1984). Perhaps the best approach to understanding debris flow is in the context of the other types of sediment-water flow.

Pierson and Costa (1987) presented a concise semi-quantitative classification scheme for types of sediment-water flow based principally on two rheologic properties: velocity and sediment concentration (Fig. 13). In this scheme, the varieties of sediment-water flow are grouped in two major categories, distinguished by sediment concentration levels and, hence, mechanical behavior: (1) apparent liquid flow, which includes normal streamflow and hyperconcentrated streamflow; and (2) plastic fluid flow, which includes slurry flow (synonymous with debris flow *sensu* Pierson and Costa, 1987) and granular flow. In both slurry (debris) flow and granular flow, interparticle mechanics are

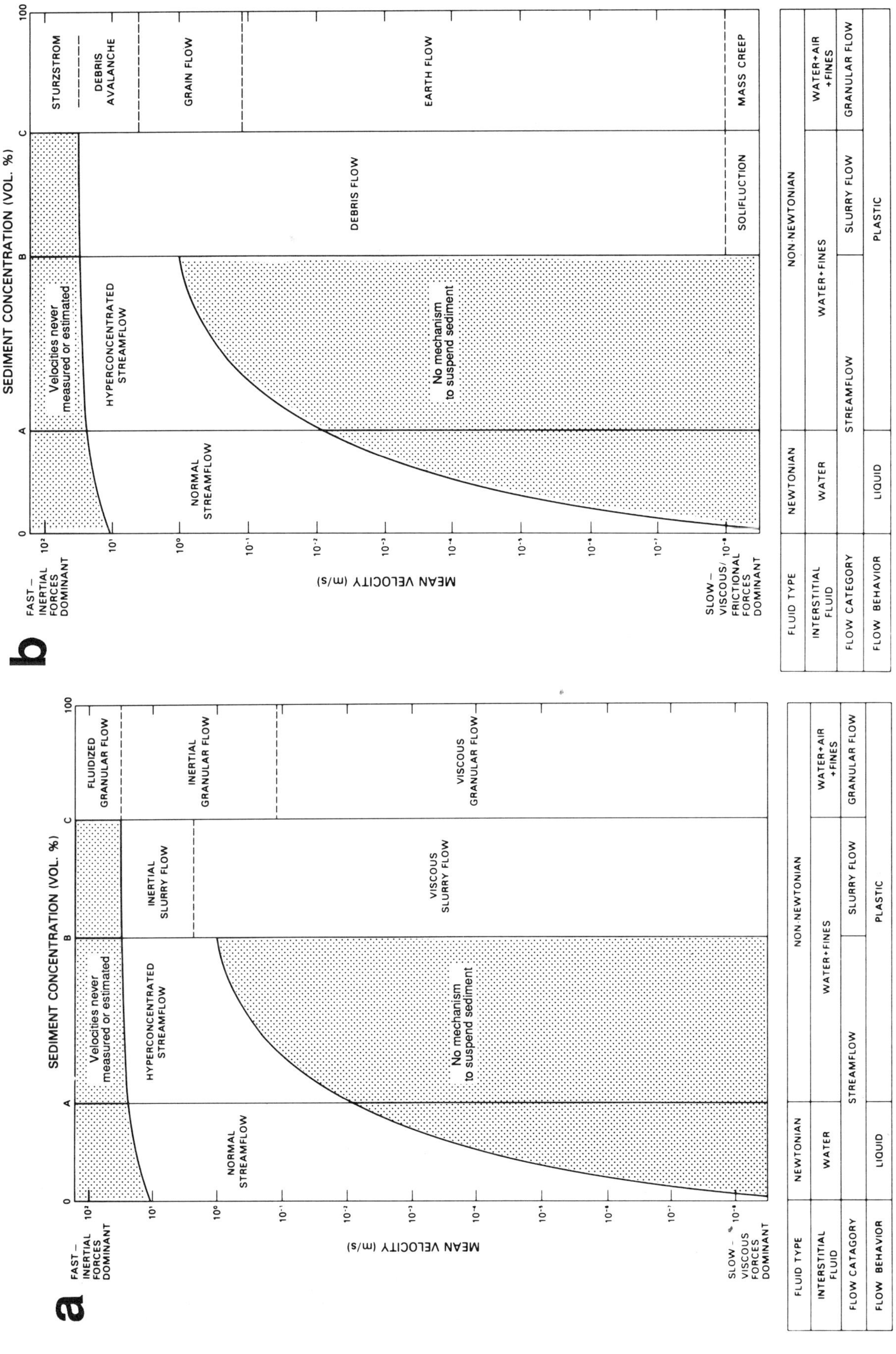

Figure 13. Classification of sediment-water flows (after Pierson and Costa, 1987). (a) Nomenclature proposed by Pierson and Costa (1987). (b) Equivalent processes in preexisting nomenclature.

dominated by viscous forces at low velocity, and by inertial forces at higher velocity. In this definition of debris flow, velocity may range over more than nine orders of magnitude (from $\sim 10^{-8}$ to $>10^{1}$ m/s), in contrast to previous definitions restricted to only high velocities (Pierson, 1984).

The transitions between normal streamflow, hyperconcentrated streamflow, slurry flow, and granular flow correspond to important rheologic thresholds crossed as sediment concentrations increase. Normal streamflow exhibits Newtonian behavior, unaffected by sediment load (Pierson and Costa, 1987). When the concentration of suspended sediment increases sufficiently to give the fluid a measurable (but very low) yield strength, streamflow is hyperconcentrated and turbulence is noticeably diminished. Technically, the onset of hyperconcentrated streamflow is the lower threshold of plastic behavior, because the fluid processes some yield strength. However, in practical terms, the yield strength of hyperconcentrated streamflow is so small that the mixture still appears to flow as a liquid. Hence, Pierson and Costa (1987) define the lower limit of slurry flow (the onset of plastic fluid flow) as the concentration that gives the fluid enough yield strength to suspend gravel-size (or larger) particles and form steep, lobate fronts and lateral levees. The debris flow mixture is saturated and is at least partially liquefied; i.e., pore fluids pressurized beyond the hydrostatic condition bear part of the weight of the solid phase. Finally, mixtures with extremely high sediment concentrations deform by granular flow. Liquefaction is impossible in this state, either because the pores are not saturated, or because high permeability allows incipient excess pore pressures to be readily dissipated.

As mentioned above, inertial forces dominate grain interactions in high-velocity flow. The transfer of momentum between colliding particles in such a medium creates dispersive pressures capable of separating, and hence supporting, large particles (Bagnold, 1954). This so-called "Bagnold effect" has been widely cited as a means of producing the peculiar inverse grading of debris flow deposits, and also as a mechanism for reducing interparticle friction, allowing long runout distances of large debris flows and avalanches even with very coarse material. The closely related process of acoustic fluidization (Melosh, 1979, 1987) is a slightly lower-energy dispersive grain-flow effect, in which dilatancy (volume increase under compressional stresses due to opening of extensional microcracks) without the loss of grain-to-grain contact is induced by acoustic waves, and achieves a similar friction-reducing effect.

Barnes (1985) suggested that the chaotic sandy breccia mapped here as the Twtb unit was emplaced by regional gravity-sliding, despite pervasive evidence favoring complex multidirectional flow rather than highly oriented sliding. A marginally better explanation of this unit's widespread distribution, sedimentology, and internal structural character employs high-velocity slurry flow or granular flow instead of gravity-sliding. However, the best interpretation of this unit's *present* internal and external structural relations probably involves no significant lateral transport at all. On the contrary, the Twtb unit exhibits much of the best evidence observed by the writer in support of in situ post-depositional disruption by seismic liquefaction. Even so, it remains likely that the *initial* deposition of the Twtb unit was by long-runout, sandy (matrix-rich), laharic debris flows.

Pyroclastic units

Air-fall tuffs, welded rhyolitic tuffs, and vulcanian breccias and agglomerates constitute the predominant pyroclastic rock types of the AVS in the study area.

Air-fall tuff. Vitric (pumice) ash-lapilli tuff is the most abundant variety of air-fall deposit, but crystal-vitric tuff, crystal-lithic-vitric tuff, and lithic tuffs are not uncommon. Euhedral biotite, euhedral to subhedral feldspars, and subhedral quartz are followed by amphibole as the most typical crystal components. All pumice fragments are at least partially devitrified.

The "Blue Point marker" (Sundell and Eaton, 1982), the vitric ash-lapilli tuff of the Blue Point beds, is among the most widespread units of the AVS, and is distributed throughout the central and southern Absaroka Range south of Wapiti Ridge (Plate IV). This marker unit varies from 0 to ⩾200 ft thick and is easily recognized by recessive, commonly slumped bentonitic slopes, which provide an invaluable reference for regional correlation (Sundell and Eaton, 1982). The vent area of the Blue Point marker is not known, although eruption from the Sunlight cone seems most likely. If this unit was ever present in the region north of the North Fork Shoshone River, it has since been removed by erosion.

Ash-flow tuff. A single thin (0–20 ft) rhyodacitic ash-flow tuff represents the basal unit of the AVS in the study area, and is recognized at scattered localities from the Jim Creek–Jim Mountain area in the North Fork Shoshone Valley to the north end of Deer Mountain in the Greybull Valley. This widespread ignimbrite is dominated by grayish glassy layers that contain various quantities of fiamme (dark, completely collapsed pumice fragments); lighter, partially collapsed pumice lapilli; broken lithic and crystal fragments; and very thin, discontinuous films of charred organic material (specimen UW1837-8). Typically overlying this welded unit are 1 to 10 ft of extremely dense, angular, finely laminated, white tuffaceous siltstones and fine-grained sandstones that contain deformed crossbeds, mm- to cm-scale climbing ripples, and fluid-escape pipes (specimen UW1837-9). Deposition of these fine sediments probably coincides with penecontemporaneous reworking of the ash-flow deposit's poorly welded uppermost layer.

Pyroclastic breccia. Vent and proximal facies mafic pyroclastic breccias and agglomerates are rare in the study area, but they form part of the middle Wapiti Formation (Twmb unit). These scarce, grayish red, grayish red purple, and light brownish gray porphyritic andesites are locally interbedded with epiclastic andesitic laharic breccias in a southward-tapering wedge that pinches out beneath Carter Mountain. More commonly, rocks of this description are found as clasts ⩽15 ft in length (specimen UW1837-10) in the Twmb unit laharic breccias, which probably originated in the Sunlight vent complex where this lithologic type is more abundant.

Pyroclastic vent breccias of felsic to intermediate composition make up much of the Wiggins Formation, erupted from the Kirwin intrusive-extrusive center south of the study area during late middle to late Eocene time (between ~47.1 and 38 Ma) (Barnes, 1985, Love and others, 1976). These angular breccias are composed principally of light grayish green, off-white, and light grayish purple fragments of rhyolite and andesite embedded in an andesite matrix (specimen UW1837-11) (Wilson, 1963, 1964b; Barnes, 1985b).

Lava flows and minor intrusions

The map distributions of lavas, sills, laccoliths, and dikes (Plates I through III) provide direct information about the loci of igneous activity, which corresponds well with the stratigraphic record of the proximal-distal volcanic facies. During Wapiti time, lava flows from the growing Sunlight stratovolcano covered an increasingly large area, while radial dikes and minor satellite intrusions were emplaced at increasing distances from the main igneous center.

Lava flows. Absaroka lavas have been referred to a suite of alkalic "basaltic" rock types—absarokite, shoshonite, banakite (Hague and others, 1899), trachyandesite, and trachybasalt (Smedes and Prostka, 1972)—all characterized by relatively high potassium and/or sodium content. In virtually all cases, the rocks are porphyritic, containing olivine, augite, and plagioclase phenocrysts in various proportions according to their precise rock type. Nearly all of these mafic lavas contain magnetite (or possibly ilmenite) in sufficient abundance to render occasional instrumental problems during mapping. Less-common flows of porphyritic dacite contain hornblende and plagioclase phenocrysts. These flows are interbedded with the basaltic lavas. Structureless autoclastic flow breccias (autobreccias), characterized by dense, monolithologic, mafic fragments, occur in subequal proportion with columnar lavas, which include both densely crystalline and vesicular/amygdaloidal textures.

Some of the AVS lava units are remarkable in their lateral persistence. For example, one flow of the upper Wapiti Formation (Twuf unit; formerly included with the Trout Peak Trachyandesite [Nelson and Pierce, 1968; Pierce, 1970]) extends as a fairly uniform sheet from its presumed source in the Sunlight area to several miles south of the Greybull River, a distance of roughly 55 mi (90 km). This laterally persistent flow is useful in correlating the top of the Wapiti Formation across the broad valley of the Greybull River, but it is indistinguishable among the numerous flows of identical composition north of Carter Mountain. This lava sequence (Twuf unit) of the upper Wapiti Formation merges to the northwest with the overlying Trout Peak Trachyandesite, increasing both in number of flows and total thickness toward the Sunlight source area. The Jim Mountain Member of the Wapiti Formation (Nelson and Pierce, 1968) is far less continuous, and thins southward from a sequence 1000 ft thick (300 m) at the Jim Mountain type section to only one discontinuous flow locally in the Hardpan Creek–Houlihan Creek area of the South Fork Shoshone Valley.

Concordant intrusions. Several small sills and one possible laccolith have been identified in the study area. These include mostly felsic and some mafic intrusions, typically less than 100 ft (30 m) in thickness and as much as 1,000 ft (30 m) in the exposed horizontal dimension. The sills appear to be comagmatic with larger felsic intrusions in the region, such as the "quartz-diorite-porphyry" (quartz-dacite-porphyry?) Deer Creek intrusion (~UTM 12TXD0891) (Rouse, 1933), and the late middle or late Eocene dacitic stock immediately southwest of Francs Peak (~UTM 12TXD6731).

The single laccolith tentatively identified within the study area is inaccessibly exposed high on the cliffs of Houlihan Creek Canyon (UTM 12TXE184076). It is clearly columnar and probably either trachyandesitic or trachybasaltic. Whereas this body exhibits a bulbous, lenticular cross section, its floor is only partially planar and lies slightly oblique to the bedding in the underlying epiclastic sequence, in contrast to the concordance expected of ideal laccoliths. Moreover, this body lies at the same stratigraphic level as the discontinuous flows of the Jim Mountain Member of the Wapiti Formation (Twj unit), and thus it may simply represent a geometrically unusual surface flow.

Discordant intrusions. Dikes are the only abundant variety of discordant intrusion within the study area. Most of these are andesite or trachyandesite, but in places dacite dikes constitute a majority. Phenocryst phases in mafic dikes include olivine, augite, and plagioclase, whereas the felsic dikes are dominated by plagioclase laths and well-oriented hornblende needles.

In most cases, radial patterns indicate clearly the source of these dikes; the most prominent pattern developed within the map area converges on the dacitic intrusion southwest of Francs Peak and closely related intrusions in that vicinity. The internal structure of some dikes clearly indicates formation by multiple injection. Subhorizontal columnar jointing structure commonly is developed, but in a few cases, columns are complexly oriented in a range of attitudes between horizontal and vertical, suggesting local perturbations of isothermal surfaces during cooling.

Contact relations between dike and country rock vary widely across the field area. In most cases, both walls are planar and bounded by a resistant baked and perhaps silicified zone as much as 2 ft wide, but in a limited number of examples, the walls are irregular, with bulbous embayments of dike rock into country rock alternating with cusps of pinched country rock.

Next to each planar dike, the altered wall rocks are extensively sheared and have steeply pitching (70 to 90°) slickenside striations on closely spaced (0.5 to 5 cm), highly polished fault surfaces. The kinematic significance of these subvertically sheared dike margins is not clear, but it is certain that the indurated rocks of these contact zones would be much more apt to record frictional differential movements than the adjacent poorly indurated epiclastic sediments if the entire sequence experienced widespread vertical motions.

The interpretation of the less common bulbous dike contacts is less speculative. Clearly these dikes intruded into unlithified, probably saturated sediment, despite burial depths of as much as several hundred feet or more. The irregular geometry of the dike walls indicates that: (1) the intruded medium was incapable of sustaining open fractures, and hence, the magma must have made room for itself as it migrated; and (2) there existed a ductility contrast between the dike and the intruded medium, with the more viscous magma pinching the more ductile soft sediments in tight cusps (specimen UW 1837-13). These sediments' ability to flow as a liquid despite their burial at depths of hundreds of feet is significant, and is examined in following sections in the context of seismic liquefaction.

Other varieties of discordant intrusions are rare in the study area. Intrusive bodies of appreciable size that are closest to the study area are the dacitic-granodioritic stocks in Meadow Creek Basin (UTM 12TXD3765) and southwest of Francs Peak (UTM 12TXD3167). A few small igneous bodies scattered across the field area may represent intrusive plugs, or they may be erosional remnants of lava flows. Examples include the pinnacle of vesicular mafic breccia just southwest of Houlihan Creek (UTM 12TXE190061) and the columnar-jointed, porphyritic intermediate-composition lavas that cap the 11,701-ft-elevation peak southwest of Timber Creek Mountain (UTM 12TXD418700).

Description and correlation of AVS units

Plate IV illustrates the stratigraphic relations among the more significant AVS units and informal subunits in a ~35-mi traverse from NNW to SSE through the study area in the east-central Absaroka Range. This cross section is constructed from the map data of Plates I through III, and recognizes a number of informal, but regionally meaningful, subdivisions of units represented by the formal nomenclature of the AVS. Several additional map units recognized in this study are not exposed along the traverse of Plate IV. Descriptions of all AVS units mapped in Plates I through III appear in sequence on Plate V, and Tables 1 and 2 summarize the genetic types, volcanic facies, lithofacies, sedimentary structures, and depositional interpretations of these units.

The value of the Blue Point marker tuff (Tbp unit) for regional correlation is readily apparent in Plate IV, as are the relations between the Tbp unit and overlying units. Because the Blue Point beds are overlain in the southern reaches of the study area by the Tepee Trail Formation, and elsewhere by the Wiggins Formation, it seems appropriate to abandon Wilson's (1963) usage—"Blue Point Member" of the Wiggins Formation—and refrain from assigning this section of air-fall tuffs, bentonite beds, blue-green lithic-arkosic sandstones and andesitic conglomerates to any existing formation. This opinion is shared by Bown (1982), Sundell and Eaton (1982), and Eaton (1982). These strata are referred to here simply as the Blue Point beds (Fig. 4); further study of regional stratigraphic relations may justify elevating these beds to the rank of a separate formation.

The Cathedral Cliffs Formation (Pierce, 1963b) comprises the mainly light-colored, massive-layered pyroclastic and epiclastic rocks known previously as the Early Acid Breccia (Hague, 1899). It is a unit of the proximal facies and represents the earliest record of Eocene volcanism in the northeastern Absaroka Range. Exposures of the Cathedral Cliffs Formation are limited and scattered, because of their involvement in Heart Mountain faulting (Pierce, 1963b) and subsequent burial by younger AVS rocks (Hague, 1899; Pierce, 1963b). Although no rocks in the east-central Absaroka Range precisely fit the descriptions of the Cathedral Cliffs Formation or the Early Acid Breccia, and while none have been mapped by either of these names, it is proposed here that the lower AVS strata in the study area may be stratigraphically equivalent. Pierce (1963b) attributed the Cathedral Cliffs Formation to a vent area in the northernmost Absaroka Range (southwestern Montana), and the dominantly epiclastic, tuffaceous lower Wapiti/Aycross rocks (particularly the Twls unit) farther south may well represent the medial to distal facies deposit of essentially the same age.

Biostratigraphic data from the Cathedral Cliffs Formations are sparse, but flora from a small collection by Pierce (1963b) indicate that it was depsited in late early Eocene or early middle Eocene time. Dicotyledon specimens from this collection suggest a more specific early middle Eocene (early Bridgerian) age (Pierce, 1986b, p. 16). This age corresponds closely with that reported by Eaton (1982) and Torres (1985) for the lower Wapiti/Aycross section. Wilson (1963, p. 14) espoused the potential correlation of the same rocks, although his remarks were couched in now-abandoned stratigraphic nomenclature (Pitchfork Formation and Early Acid Breccia).

Torres and Gingerich (1983) and Torres (1985) assert that in the North Fork Shoshone River Valley, Paleozoic limestones were emplaced by Heart Mountain faulting onto early-middle Bridgerian fluvial volcaniclastic rocks, which they called Aycross Formation. Pierce (1986) ascribes these lower plate rocks to the Cathedral Cliffs Formation, in keeping with his definition of that unit as the only volcanogenic unit pre-dating Heart Mountain faulting. Because the Aycross Formation is generally restricted to the more distal environment of the southeastern Absarokas, these early-middle Bridgerian lower plate rocks are better considered as lowermost Wapiti Formation. In either case, both the pyroclastic Cathedral Cliffs Formation and the fluvial epiclastic rocks of the lowermost AVS in the east-central Absaroka Range (lower Wapiti/Aycross strata) may well be stratigraphically equivalent and older than Heart Mountain faulting. This probable equivalence demonstrates the importance of understanding the proximal-distal facies relations illustrated in Figures 4 and 5, and the hazards of basing regional correlations on diachronous textural boundaries in volcanic-volcaniclastic terranes. In addition, the age equivalence suggested by this correlation increases the likelihood of a genetic relation between Heart Mountain faulting and the liquefaction-related deformation of the lower Wapiti and Aycross Formations in the study area.

In the northern Teton Range, the Hominy Peak Formation

TABLE 1. PRINCIPAL GENETIC TYPES, VOLCANIC FACIES, AND LITHOFACIES OF AVS UNITS IN THE EAST-CENTRAL ABSAROKA RANGE.

Map Unit	Principal Genetic Type(s)	Volcanic Facies*	Principal Lithofacies Code(s)†
Twi	pyro-, epiclastic	V-M	Gms, Gmu, Gt, Gp, St, Sp
Ttwi	pyro-, epiclastic	V-M	transitional between Twi & Tt
Tt	epiclastic	P-M	Gms, Gmu, Gm, Gp, Sp, Sh
Tbp	pyro-, epiclastic	P-D	Gt, Gp, St, Sp, Sh
Ttp	lava flow	V-P	columnar §
Twuf	lava flow	V-M	columnar > autobrecciated §
Twu	pyro-, epiclastic	M	Sh, Fh, Fla, C
Twmb	epiclastic	P	Gms, Gmu, Gm, Gt, Gp, Sm, St, Sp
Twj	lava flow	V-P	columnar > autobrecciated §
Twls	epi-, pyroclastic	M-D	Sp, Sh, Fh, Fla
Twf/Twfb	lava flow	V-M	autobrecciated > columnar §
Twtb	epiclastic	P-M	Gms, Gm
Tws	epiclastic	M	Sp, Sh, Gh, Fla
Twca	epiclastic	M	Gmu, Gm, Gt, Gp, Sh
Twlb	epiclastic	P-M	Gmu, Gms, Gm
Tw	epiclastic	M	variable (all?)
Tas	epiclastic	M	Sm, Sh, Sp, Fh
Ta	epi-, pyroclastic	M-D	Gm, Gt, Sh, Fh, Fla, C

* Volcanic facies: V = vent, P = proximal, M = medial, D = distal; dash indicates intervening facies are also represented.

† Lithofacies codes adoped with addition of Sm from Fritz and Harrison (1985). Refer to Table 2 (modified from Fritz and Harrison, 1985) for explanation of codes, interpretation of process, and depositional environment. Truly pyroclastic deposits are largely excepted from this lithofacies scheme.

§ Lava flow units are of typical Absaroka compositions (i.e., trachyandesite + trachybasalt + absarokite + shoshonite >> dacite).

(Love and others, 1978) is the oldest AVS unit, and like the Wapiti and Aycross Formations in the study area, its basal unit is a white marker tuff. It is possible, although not yet provable, that these widely separated tuffs are equivalent remnants of a once-continuous pyroclastic blanket. In any case, their approximate equivalence over so broad an area testifies to the regional significance of earliest Absaroka siliceous (explosive) volcanism.

Summary

The major lithologic elements of the AVS are of three essential types: (1) lavas; (2) pyroclastic deposits; and (3) epiclastic volcanic deposits derived directly from these primary volcanic rocks, commonly with very brief pre-depositional sedimentary histories. In actuality, the subtleties of the AVS's volcanic-volcaniclastic stratigraphy are far more complex than can be expressed here. Comprehensive modern investigations of AVS lithofacies and depositional processes in this area would complement recent research in the northern (Fritz and Harrison, 1985) and southern (Bown, 1982, b; Sundell, 1982; Barnes, 1985) parts of the Absaroka Range, and other Cenozoic intermediate-composition volcanic fields as well (Smith, 1986; numerous references cited therein). Similarly, continued refinements of AVS correlation will further clarify the temporal relation to Heart Mountain faulting, and perhaps help to illuminate the fundamental cause of early middle Eocene deformation.

STRUCTURAL STYLE AND KINEMATICS OF EARLY MIDDLE EOCENE DEFORMATION IN THE WAPITI AND AYCROSS FORMATIONS

Introduction

An exceptionally broad range of styles and orientations characterizes the deformational structures in the lowermost units

TABLE 2. LITHOFACIES CODES AND SEDIMENTARY STRUCTURES IN VOLCANICLASTIC SEDIMENTARY ROCKS COMMONLY ASSOCIATED WITH CONTINENTAL INTERMEDIATE-COMPOSITION VOLCANIC FIELDS*

Facies Code	Lithofacies	Sedimentary Structures	Interpretation
Gms	massive matrix-supported gravel	massive bedded	debris flow
Gmu	massive, unsorted gravel	normal-inverse graded crudely bedded, no fabric or imbrication	debris flow
Gm	massive or crudely bedded gravel	horizontal bedding, imbrication	hyperconcentrated stream flow, sieve & lag deposits, bars, sediment flow
Gt	gravel, stratified	trough crossbeds	channel fill
Gp	gravel, stratified	planar crossbeds	linguoid bars or deltaic growth from older bar remnants
Sm	medium- to very coarse-grained sand, may be pebbly, mostly volcanic grains	massive, structureless	sandy debris flow or post-depositional disruption of strata by liquefaction
St	"	solitary or grouped trough crossbeds	lower flow regime dunes
Sp	"	solitary or grouped planar crossbeds	linguoid, transverse bars, sand waves, lower flow regime
Sh	very fine- to coarse-grained sand, pebbles	horizontal lamination	planar bed flow, upper and lower flow regimes
Fh	silt, mud, ash	crude horizontal bedding	planar bed flow, traction carpets
Fla	sand, silt, mud, ash	fine lamination, very small ripples	overbank, waning flood deposits
C	coal, carbonaceous mud	plant, mud films, paleosols, organic zones	swamp, lacustrine overbank deposits

* The first (capital) letter of each facies code refers to grain size and additional letters refer to stratification. This table modified from Fritz and Harrison (1985) with addition of Sm code entries. Remaining lithofacies codes modified from Miall (1978) with addition of Gmu, Fh, and Fla codes from Harrison (1985) by Fritz and Harrison, (1985).

of the AVS in the east-central Absaroka Range. Stratigraphic relations constrain this deformation to the interval of early middle Eocene time immediately following the deposition of laterally equivalent sections of the lower Wapiti and Aycross Formations. The rootless style and nonuniform kinematic (movement) patterns of structural development appear to show that gravitational or density instability was the principal driving mechanism for many of the structures created by this deformation. However, a satisfactory explanation of what created the numerous local gravitational instabilities across the area is lacking from previous investigations.

Deformation at the intra-Wapiti/Aycross stratigraphic level is widely recognized throughout the study area, but it is most intense in local zones, referred to as domains, typically dominated by groups of stylistically similar structures. These domains are categorized into four principal types (Decker, 1987) on the basis of their dominant structural style (Fig. 14):

Type 1. Tight to isoclinal overturned folds.

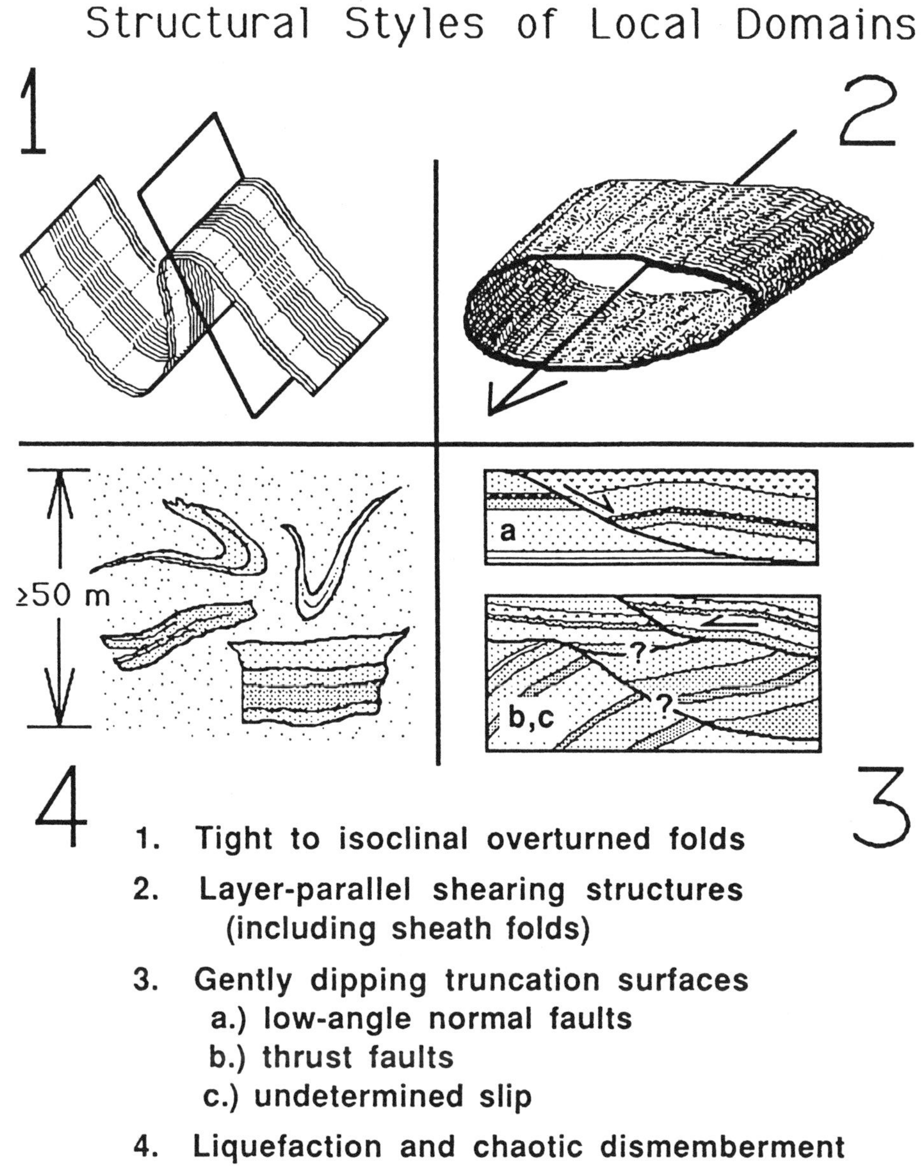

Figure 14. Schematic representations of structural style domain, Types 1 through 4. The flattened cylinder in the upper right compartment represents part of a bedding surface in a sheath fold, whose axis (arrow) lies parallel (or antiparallel) to the direction of shear transport.

Type 2. Layer-parallel shearing structures, including sheath folds.

Type 3. Gently dipping truncation surfaces; (a) low-angle normal faults, (b) thrust faults, and (c) faults of undetermined direction of slip.

Type 4. Chaotic dismemberment.

The applications and limitations of this classification scheme are discussed at the end of this section. The distribution of domains of Types 1 through 4 in the east-central Absaroka Range is shown in Figure 15.

Type 1 domains: Tight to isoclinal overturned folds

The best-exposed Type 1 domain is the southeast-verging fold train (Fig. 16a) located on the east side of Deer Mountain in the Timber Creek drainage (UTM 12TXD437760). This domain comprises a set of predominantly tight, smooth, cylindrical folds whose interlimb angles range from 0° (isoclinal) to about 110° (open). They have amplitudes of about 30 to 200 ft, wavelengths of about 30 to 300 ft, and they die out at their lower enveloping surface. Folding is disharmonic between upper and lower parts of the folded sequence. Smaller folds occur in the thinner, less competent beds near the top. Type 1 folds are most readily recognized in well-bedded intervals, and similar deformation may go unnoticed in massive deposits, such as the laharic breccias so prominent in these formations.

Type 1 domains are not common in the study area, and their structure and orientations are highly inconsistent from one to the next. The southeast vergence of the Timber Creek folds stands in

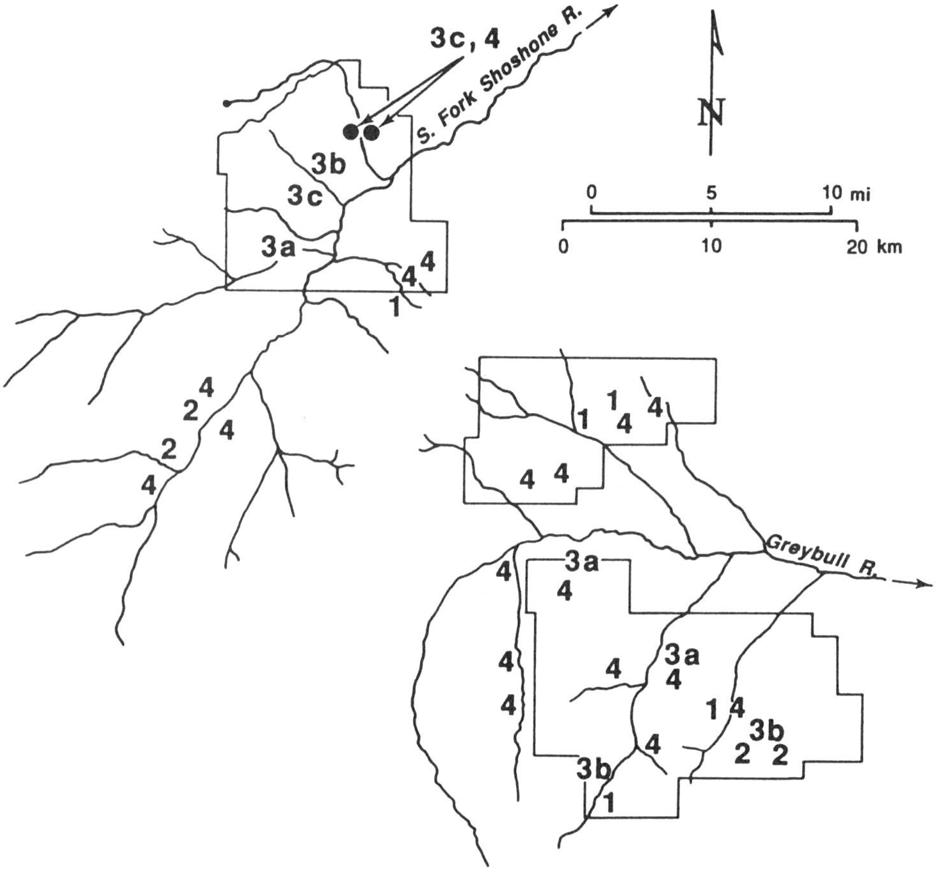

Figure 15. Geographic distribution of local domains of Types 1 through 4.

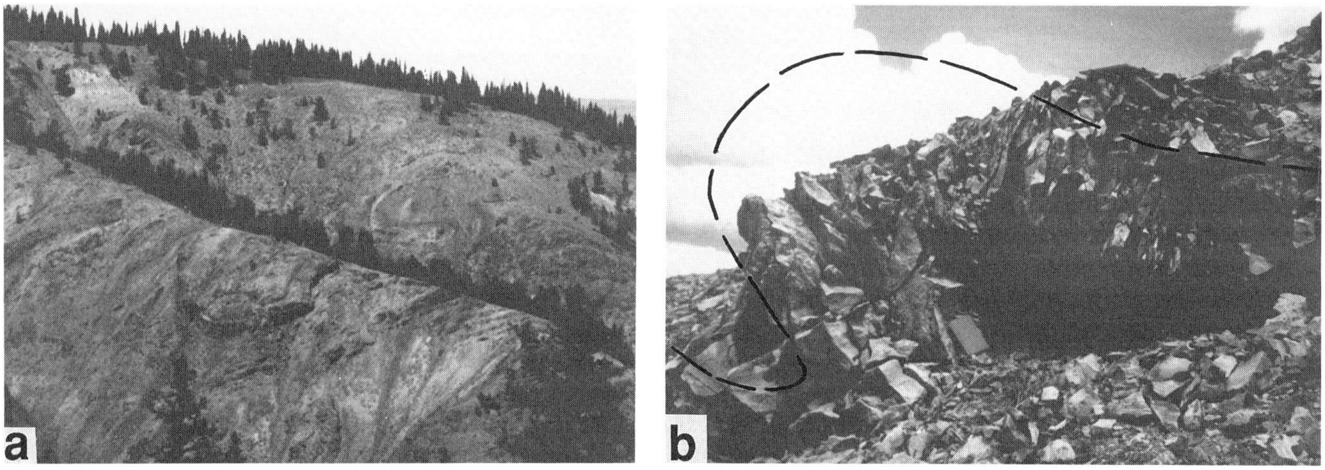

Figure 16. Photos of rounded Type 1 folds in Aycross Formation. (a) View to northeast of southeast-verging fold train 1 mi west of Timber Creek at ~UTM 12TXD437760. (b) Smaller WNW–verging folds in pale greenish yellow, laminated dust tuff on north end of Deer Mountain at ~UTM 12TXD422783, view to ~north. Field notebook lies parallel to axial plane of anticline-syncline pair. Folding strain was accommodated here by intense transposition on closely spaced microfaults (see Fig. 37) before the muddy tuffaceous sediments were completely lithified.

marked contrast to the WNW vergence of smaller Type 1 folds only ~1.7 mi to the NNW (Fig. 16b) on the north end of Deer Mountain (UTM 12TXD422783), and to the approximately north vergence of large box folds (Fig. 17) in upper Francs Fork (UTM 12TXD390720). Common to each, however, is an apparent local discontinuity between folded and unfolded strata at the lower enveloping surface. The rootless character of these folds suggests shortening by gravitational drive, but no discrete detachment fault or landslide surface is developed. Instead, displacement appears to have been accommodated by distributed flow in a zone of "depth creep" (Ter Stepanian, 1963, 1974). One possible scenario of Type 1 fold development incorporates this fundamental characteristic in an application of infinite-slope soil mechanics theory to the Timber Creek fold train.

Type 2 domains: Layer-parallel shearing structures

Layer-parallel shearing is indicated by a suite of structures, developed in two sets of domains at opposite ends of the study area (Fig. 15). In the eastern set, high shear strains are recorded by several well-exposed sheath folds (Figs. 18, 19), while in the western set, lower shear strains are indicated by scattered isoclinal recumbent intrafolial folds, tight drag folds, and associated gently inclined synthetic shear planes with small normal offsets (Fig. 20).

This study recognizes for the first time the occurrence of sheath folds in the AVS. These remarkably well-developed sheaths are quite unusual in the shallow regime of the AVS during early middle Eocene time; they are more commonly associated with ductile deformations under medium- to high-grade metamorphic conditions. Their concentration between Timber Creek and Dick Creek near the corner of the field area documents a southwest-trending avenue of exceptionally high shear strains, perhaps as great as $\gamma \geqslant 10$ (Cobbold and Quinquis, 1980), which in this supracrustal setting would be unachievable except in an unlithified state. The precise nature of the gravitational instability exploited to develop these strains remains unclear, although the chaotic interval in which several of them occur appears to have been dismembered during partial liquefaction. One plausible setting for strong, distributed, layer-parallel shearing would be in a zone of "depth creep" (Ter Stepanian, 1974). In such a regime, irregularities in bed thickness or other deviations from planar stratification would cause perturbations in the flow field, which at high shear strains, cause sheath folds and other passive folds to form (Cobbold and Quinquis, 1980; Hudleston, 1983). The depth creep model is explained more fully in a following section in the context of Type 1 fold development.

Although relatively few in number, Type 2 structures provide unequivocal evidence of local dominantly horizontal transport. As with Type 1 folds, the kinematics of Type 2 structures are only locally uniform. Sheath folds from the eastern domains formed in a regime with displacement of upper layers to the southwest relative to lower layers; structural geometries from the western set, however, imply a predominant top-to-NE or top-to-ENE sense of shear. Moreover, few of the transport directions discussed so far are easily related to gravitational spreading away from the topographically high Sunlight volcanic center.

Figure 17. View to southeast of large, north-verging box folds of Type 1 domain in Aycross Formation, east side of tributary valley to upper Francs Fork at UTM 12TXD390720. Additional folds from this domain are partially exposed ~1 mile northeast of this locality. For scale, roughly 1,000 ft (300 m) of relief is shown on slope.

Type 3 domains: Gently dipping truncation surfaces

The gently dipping truncation surfaces of Type 4 domains include: (a) low-angle normal faults, (b) thrust faults, and (c) some surfaces of obscure origin that may be either extensional or compressional faults and, in one case, a surface that may be either a fault or an aberrant angular unconformity.

Type 3a structures. Figure 21 shows the best-exposed example of a low-angle normal fault in the study area. This structure strikes about N15°E and dips approximately 25°NW, cutting a well-bedded section of the lower Aycross Formation east of Francs Fork on the north end of Deer Mountain (UTM 12TXD4278). The geometry of this structure is very similar to that predicted by scale models of listric normal faulting at rather

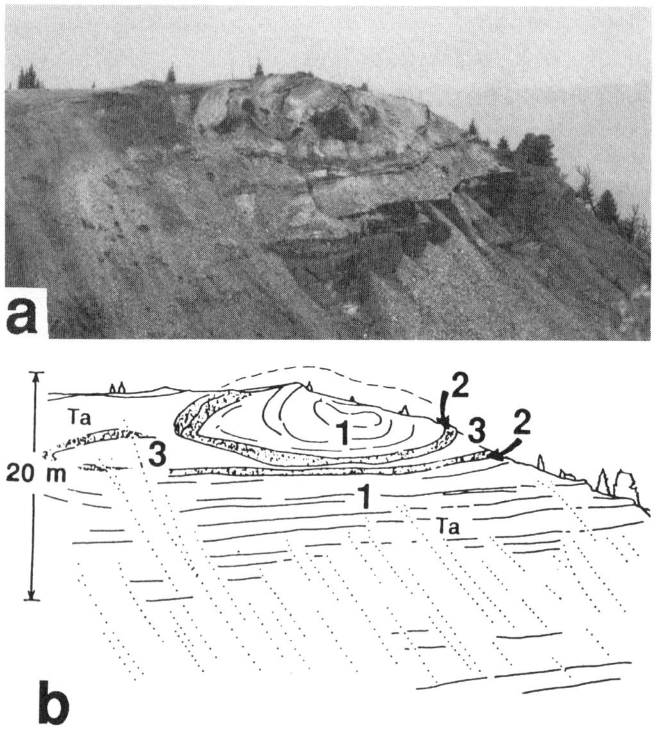

Figure 18. Sheath folds in Aycross Formation from eastern Type 2 domains, between Timber Creek and Dick Creek. (a) Photo looking ENE at flattened anticlinal sheath fold in tuffaceous beds. Sheath axis plunges gently to S42°W at UTM 12TXD475753. (b) Photosketch of fold in (a), with beds numbered (1 = oldest, 3 = youngest) to illustrate anticlinal geometry. (c) SSE–plunging anticlinal(?) sheath in volcanic wackes at UTM 12TXD470747.

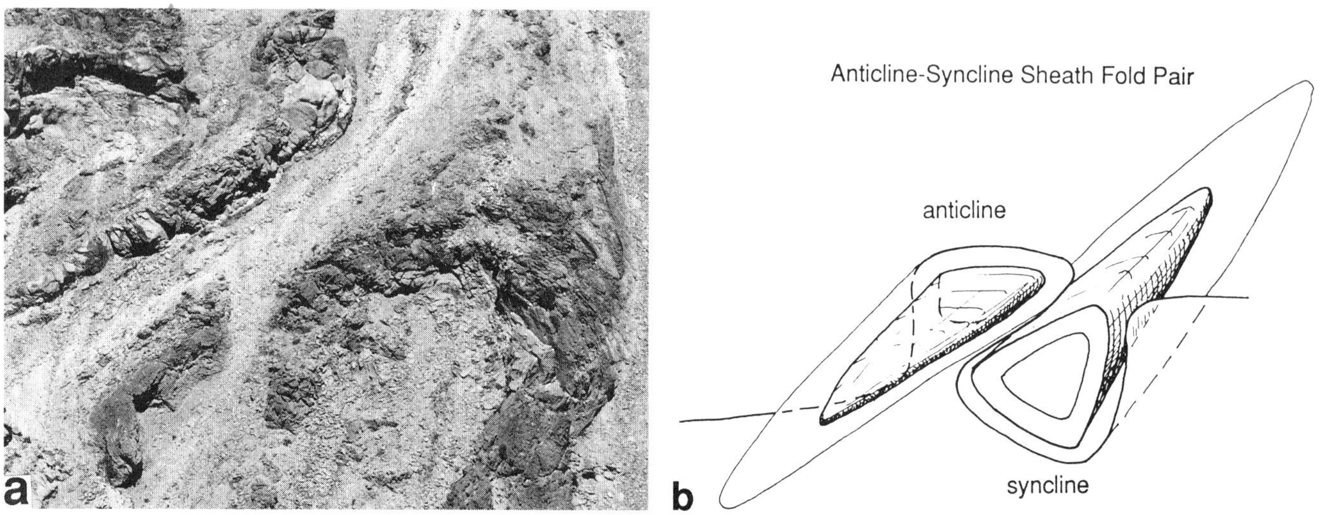

Figure 19. Photo (a) and interpretive line drawing (b) looking north at anticline–syncline sheath fold pair in Aycross Formation from eastern Type 2 domains. In anticlinal sheath folds, as in all anticlines, beds young away from the core of the fold, and in synclinal sheaths, beds young toward the core. Sheath axes plunge gently to the southwest at UTM 12TXD468749. For scale, upper left (anticlinal) sheath is roughly 65 ft (20 m) tall.

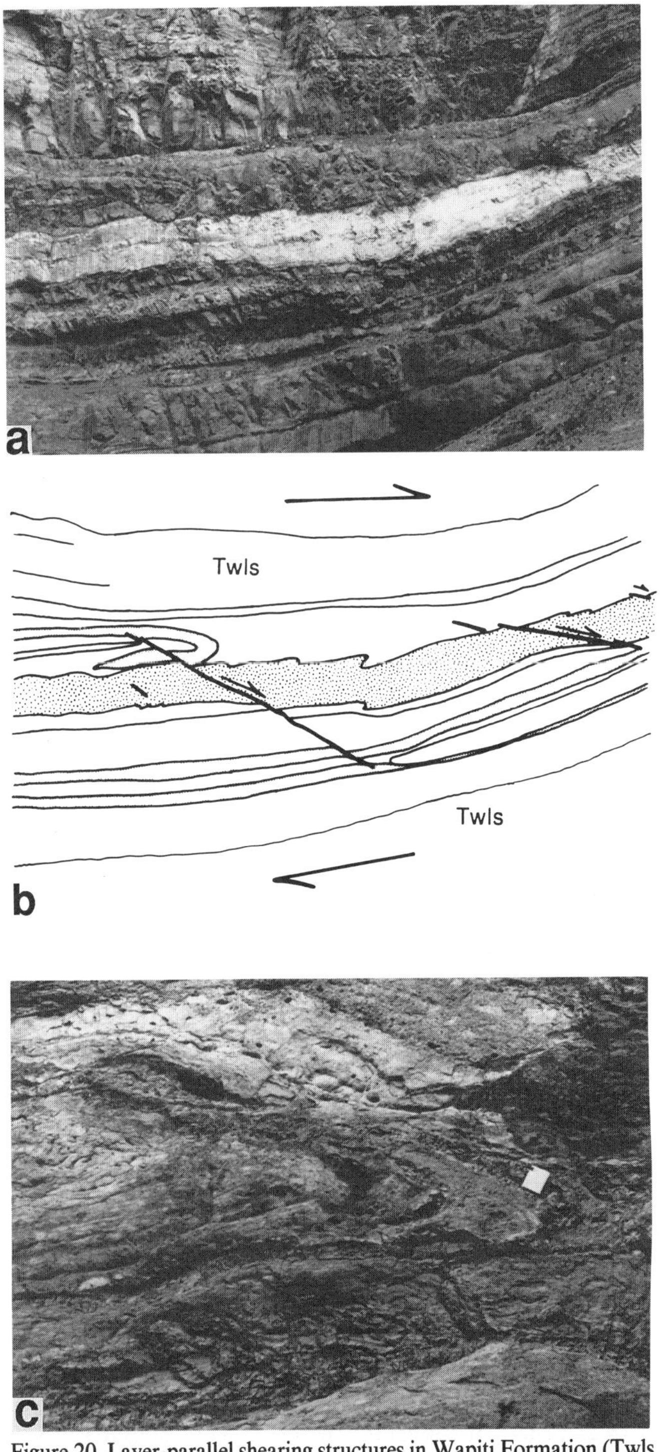

Figure 20. Layer-parallel shearing structures in Wapiti Formation (Twls unit) from western Type 2 domains. (a) Photo and (b) photosketch looking NNW at structures exposed in sheer wall near mouth of Deer Creek canyon at UTM 12TXD102906. A rootless isoclinal recumbent fold with sheared-off overturned limb, several synthetic shears, and numerous small drag folds consistently indicate top-to-NE or -ENE sense of shear. (c) View to northwest of isoclinal recumbent intrafolial fold about 1 mi SSW of Deer Creek at UTM 12TXD0989. The geometry of this and other folds in the Twls unit in the upper South Fork Shoshone valley are compatible with the top-to-NE or -ENE sense of shear shown in (a).

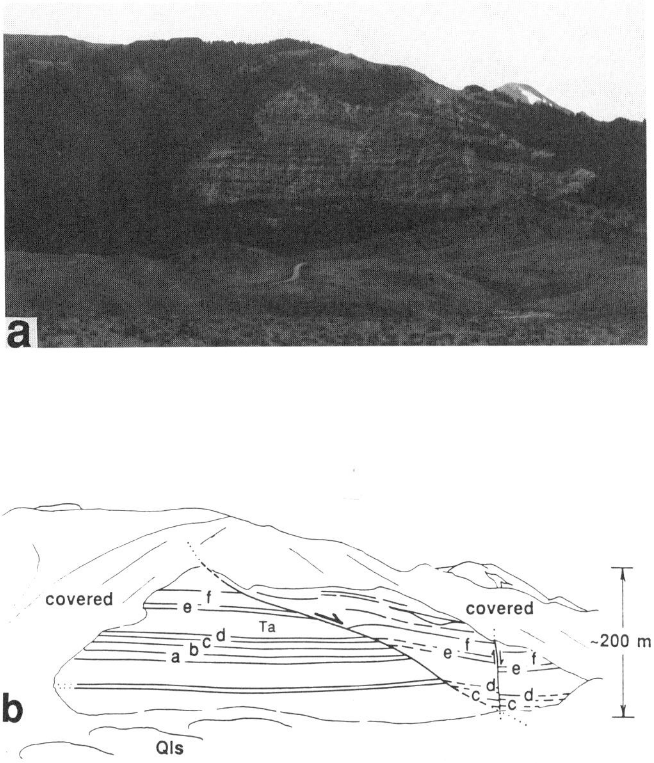

Figure 21. Photo (a) and photosketch (b) looking south at low-angle normal fault of Type 3a domain in Aycross Formation east of Francs Fork at UTM 12TXD4278. Beds are labeled in (b) for correlation across faults. Apparent lateral transport of the hanging-wall block is about 270 ft (80 m) to N75°W.

Figure 22. View to west of probable low-angle normal fault in Aycross Formation, just northwest of fault in Figure 21 at UTM 12 TXD421787. Beds cannot be matched across the gently north-dipping fault, but a nearly horizontal, crudely developed fracture cleavage (*sensu* Wilson, 1982) in the footwall adjacent to the fault suggests a normal sense of shear. Note figure near fault for scale.

low values of extensional strain (McClay and Ellis, 1987, Fig. 6, p. 343), complete with the development of a modest rollover anticline and steep antithetic faults in the hanging wall. Apparent normal dip-slip is approximately 300 ft (90 m), which implies roughly 270 ft (80 m) of horizontal transport of the hanging-wall block to ~N75°W. This fault is observed only in a single exposure, but it appears that the dip shallows at the base of this cliff, and that the contact with the underlying Willwood Formation is probably not offset. A second low-angle fault oriented N79°W, 37°NE crops out in a limited area immediately northwest of the first (~UTM 12TXD421787). The hanging-wall and footwall blocks of this fault (Fig. 22) contain no matching strata, but a 50- to 120-cm-thick gouge zone subjacent to the fault contains a crudely developed, subhorizontal fracture cleavage (*sensu* Wilson, 1982), which suggests a normal sense of shear.

A third fault, in many respects nearly identical to the first, is exposed ~6 mi to the northwest just west of Willow Creek (UTM 12TXD3585). This fault strikes roughly N25°E and dips approximately 25°NW, and shows nearly 500 ft (150 m) of dip-slip. It follows that horizontal translation of the hanging wall is approximately 450 ft (140 m) toward ~N65°W. The close geometric and kinematic relations between the Willow Creek and Francs Fork faults suggest that they belong to the same dynamic regime. Together, these listric faults accommodate a small WNW–ESE rootless extension (~2 percent minimum), but more significantly, they constitute the largest domain of consistent structural style found in the study area.

Type 3b structures. Thrust faults with small offsets are observed in a number of localities in the field area, and span a broad range of transport directions. Figure 23a depicts a thrust fault that cuts a variegated fluvial epiclastic and pyroclastic section of the Aycross Formation east of Timber Creek (UTM 12TXD478756). This thrust has a small fault-bend fold in the hanging wall, and an apparent displacement of about 60 ft (20 m) toward the ENE.

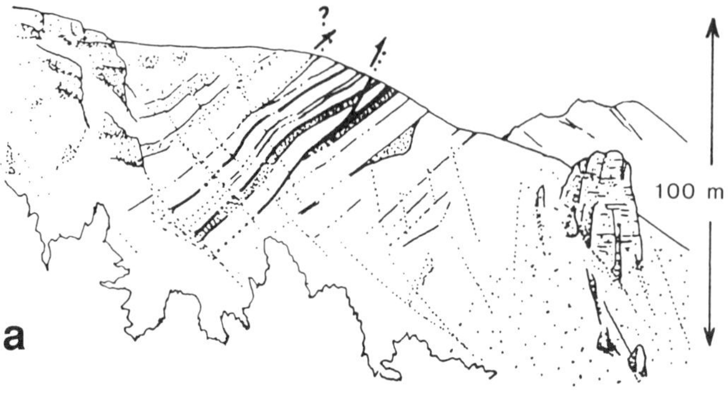

Figure 23. Small thrust faults of Type 3b domains in Aycross Formation. As with structures of all styles, inferred thrust transport directions vary widely among local domains. (a) Photosketch looking ENE at WNW–dipping thrust(s?) in variegated tuffs and fluvial epiclastic strata E of Timber Creek at UTM 12TXD478757. Apparent displacement is 65 ft (20 m) to ESE. Note minor faultbend fold above ramp, and channel profile in beds below thrust. (b) Photo looking northwest at northeast-dipping thrust faults in upper Francs Fork at UTM 12TXD392732. Apparent dip-slip displacement of bed marked by inked line is approximately 30 ft (9 m) to southwest. Incompetent mudstone bed in footwall of larger fault exhibits drag folding caused by thrust movement (between main thrust and minor synthetic fault, just left of center).

A thrust on Francs Fork (~UTM 12TXD392732) shows offset of similar magnitude, but directed toward the southwest (Fig. 23b), as do two other thrusts northeast of Houlihan Creek in the South Fork Valley (~UTM 12TXE1907). Another thrust fault between Houlihan and Hardpan Creeks is confined to the Willwood Formation (Plate I), and its ~45° NW dip suggests offset toward the southeast. This fault does not cut the overlying Wapiti Formation, but concordant bedding relations between the two formations suggest that the thrust developed simultaneously with those nearby in the lower AVS, rather than before AVS deposition.

Type 3c structures. Several truncation surfaces that defy classification on grounds of shear sense are exposed in a landslide scar on Hardpan Creek (UTM 12TXE2109 and 2210) (Fig. 24). It is clear from their geometry that most of these were produced by slip on relatively planar or curviplanar surfaces. Hay (1954) interpreted one surface on the east side of Hardpan Creek (UTM 12TXE226103) (Fig. 25, Plate VIII.B) as an angular unconformity. This surface was inspected closely in this research for evidence of its origins, but with inconclusive results. Because most of these faults were active prior to lithification, few brittle deformational effects are observed to differentiate clearly between erosional and dynamic truncation surfaces. For the same reason, few well-developed cataclastic micro- or mesofabrics are developed from which to determine the sense and precise azimuth of movement on faults where displacement is indeterminable from stratigraphic evidence alone. Exceptions to the latter include rare, and unfortunately out-of-place, hydroplastic fault surfaces (Laville and Petit, 1984; Petit and Laville, 1987).

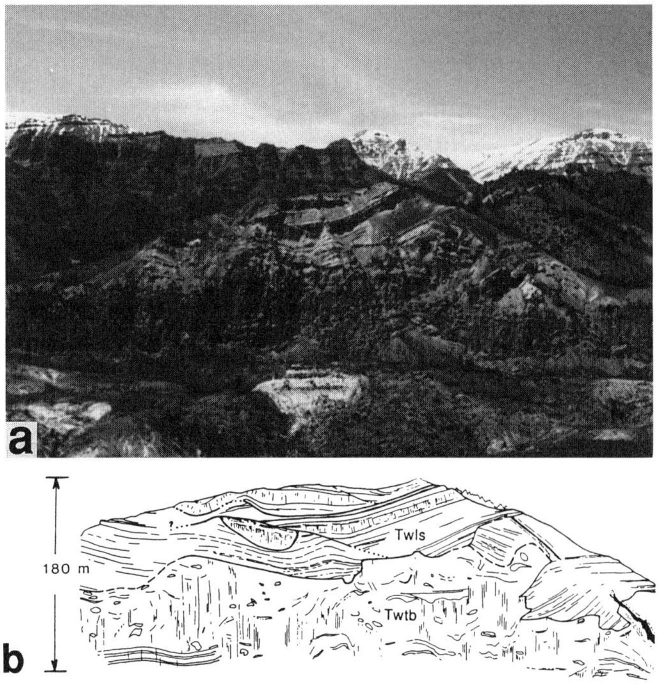

Figure 24. Photo (a) and photosketch (b) looking north at stacked Type 3c truncation surfaces in the lower well-stratified section (Twls unit) of the Wapiti Formation exposed in landslide scar just west of Hardpan Creek at UTM 12TXE2109 and 2210. These truncation surfaces are probably prelithification faults caused by liquefaction and intense dismemberment in the underlying Twtb unit, but beds cannot be matched across surfaces to ascertain sense(s) of slip.

Summary of kinematics in domains of Types 1 through 3

A common feature of the structures within domains of Types 1 through 3 is that they all bespeak local lateral transport of at least semiconsolidated strata. However, because these strata were only partially lithified during deformation, they typically contain no measurable kinematic indicators. Thus, transport directions can only be estimated from the simplest scenario justified by any particular structural geometry. For example, in the absence of other information, the vergence direction of overturned folds is taken to represent the direction of movement that caused the folding, and transport on gently inclined faults is assumed normal to fault strike, either up- or down-dip. The spectacular sheath folds discovered near the southeast margin of the study area yield the most specific kinematic data, with inferred transport parallel to the plunge of sheath axes.

Figure 26 summarizes the kinematics inferred from structures of all domains of Type 1 through 3 recognized in the east-central Absaroka Range. Its most striking feature is the lack of any uniform or radially symmetric transport pattern. This observation, coupled with the lack of throughgoing detachment surfaces and the coexistence of extensional and compressional structures, testifies to the complicated local stress regime(s) within the Absaroka volcanic-volcaniclastic pile during early middle Eocene time. Such kinematic and dynamic complexity supports a deformational model that involves the triggering of very localized gravitational instabilities, which were determined in turn by very localized physiographic and stratigraphic configurations, rather than the regional gravity-sliding (Barnes, 1985), regional gravity-spreading (Hauge, 1982, 1985), or "Absaroka Detachment Complex" (Hauge, 1986) models proposed previously. The following interpretation of Type 4 domains offers a different deformational mechanism capable of generating the inferred local instabilities and, hence, is able to integrate the area's diverse structural patterns as parts of a united whole.

Type 4 domains: Liquefaction and chaotic dismemberment

Several of the chaotic intervals classified here as Type 4 domains have received prior attention from workers who discussed some, but not all, of those analyzed here. Their interpretations ranged from chaotic deposition from massive volcanic explosions (Bucher, 1933), and interstratal injection from a volcanic vent (Hay, 1954), to various types of catastrophic surficial mass-wasting processes, including detachment faulting (Bown, 1982b), gravity-sliding (Barnes, 1985), and landsliding (J. Healey, oral communication, 1987). This study reveals numerous relations that demand critical reevaluation of these interpretations, and proposes an alternative hypothesis—seismic liquefaction—that more adequately explains these observations.

Characteristics of Type 4 domains and evaluation of previous hypotheses. The most important characteristic of Type 4 domains is the chaotic arrangement of their components. Irregularly shaped, commonly delicate inclusions are scattered in a massive matrix consisting of individual detrital grains of all size fractions from clay to small boulders. The inclusions range from a few inches to hundreds of feet across, and consist of bedded or massive volcanic siltstone, sandstone, tuff, conglomerate, breccia, or interbedded combinations of these sedimentary and pyroclastic types, and in one locality, masses of dacite. In average composi-

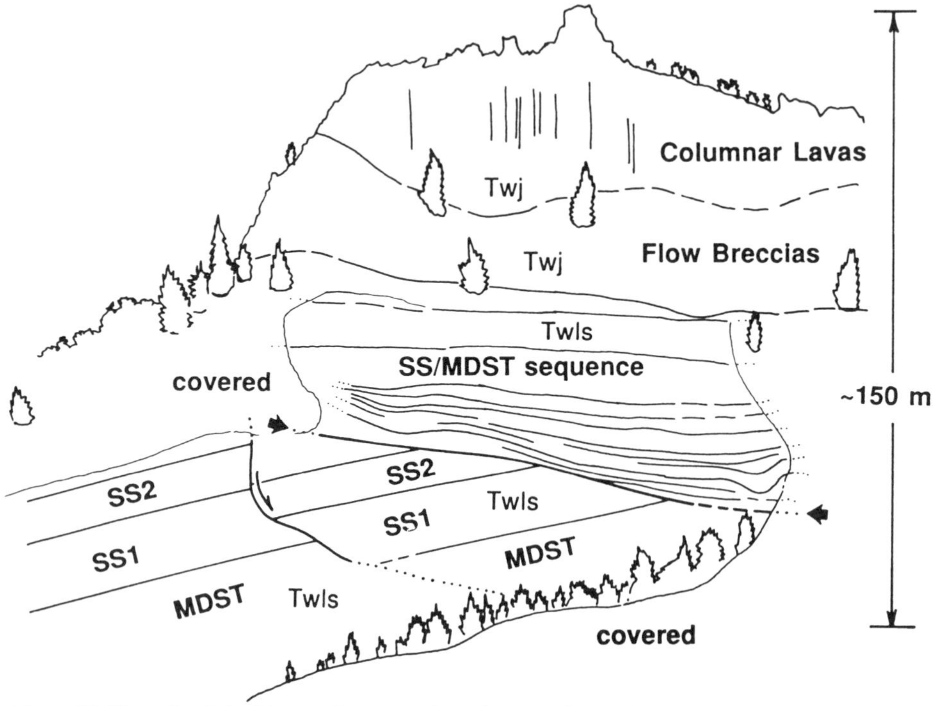

Figure 25. Photosketch looking northeast at enigmatic truncation surface (arrows) exposed in landslide scar just east of Hardpan Creek (northern map area) at UTM 12TXE226103. Hay (1954) interpreted this surface as an unconformity, but its smoothness, the truncation of beds both above and below the surface, as well as its proximity to several faults (see also Plate VII.B) and the truncations of Figure 24, suggest that it may be of structural origin. Beds of mudstone (MDST) and sandstone (SS1, SS2, etc.) within Twls unit are labeled for correlating and distinguishing beds across fault and truncation surface.

tion and color, the inclusions and matrix materials at any one locality are commonly very similar. The color and composition vary somewhat among different Type 4 domains, and in most cases, the composition closely resembles that of the laterally adjacent strata. Where exposed, the edges of Type 4 domains are typically very irregular, with complex interleaving between unbedded chaotic material and neighboring strata. Upper and lower contacts are commonly gradational, but locally are sharp, and they vary from smooth bedding-parallel surfaces, to irregular cross-cutting boundaries. The largest of these chaotic zones locally exceeds about 550 ft (170 m) in thickness, and extends laterally for more than a mile. The areal and stratigraphic distribution of Type 4 domains compares closely with the distribution of domain Types 1 through 3.

A successful model of Type 4 deformation must integrate the following characteristics interpreted from observation of these field relations:

(1) Essentially all of the bounding surfaces appear to be formed by disruption of preexisting stratified sequences rather than by deposition. Hence, deformation must have occurred in sediments after deposition, but prior to widespread lithification, with intergranular slip as the chief deformation mechanism accommodating very large strains.

(2) Deformation within chaotic intervals involved intense dismemberment and folding of some preexisting strata, and the complete obliteration and mixing of others.

(3) Deformation occurred at depths that, in places, reached at least 650 ft (200 m), with highly variable and irregular contacts between 0 to 500-ft-thick (0 to 150 m) chaotic intervals and less deformed overlying and underlying units, and laterally adjacent strata.

(4) Most deformation occurred without injection of molten lava or pyroclastic material and without substantial thermal/chemical alteration.

(5) Evidence of lateral transport is generally lacking within Type 4 domains, although in at least one case, diapiric movement of sediment had a significant horizontal component.

(6) Type 4 domains are widespread throughout the study area, and are closely associated (both temporally and spatially) with all other styles of deformation.

Many of the internal characteristics of these domains and their structural relations to adjacent strata are illustrated in Plates VII and VIII. The involvement of strata overlying and adjacent to the chaotic interval is especially significant.

In view of these characteristics, the shortcomings of each of the previously proposed interpretations become clear. Surficial mass wasting as the cause of Type 4 deformation is ruled out by observations that this deformation occurred at depth, with the involvement of overlying, relatively less deformed strata (Plates VII and VIII), and moreover, by the paucity of evidence suggesting lateral transport of material in Type 4 domains. Hypotheses limited to a chaotic volcanic depositional event are likewise un-

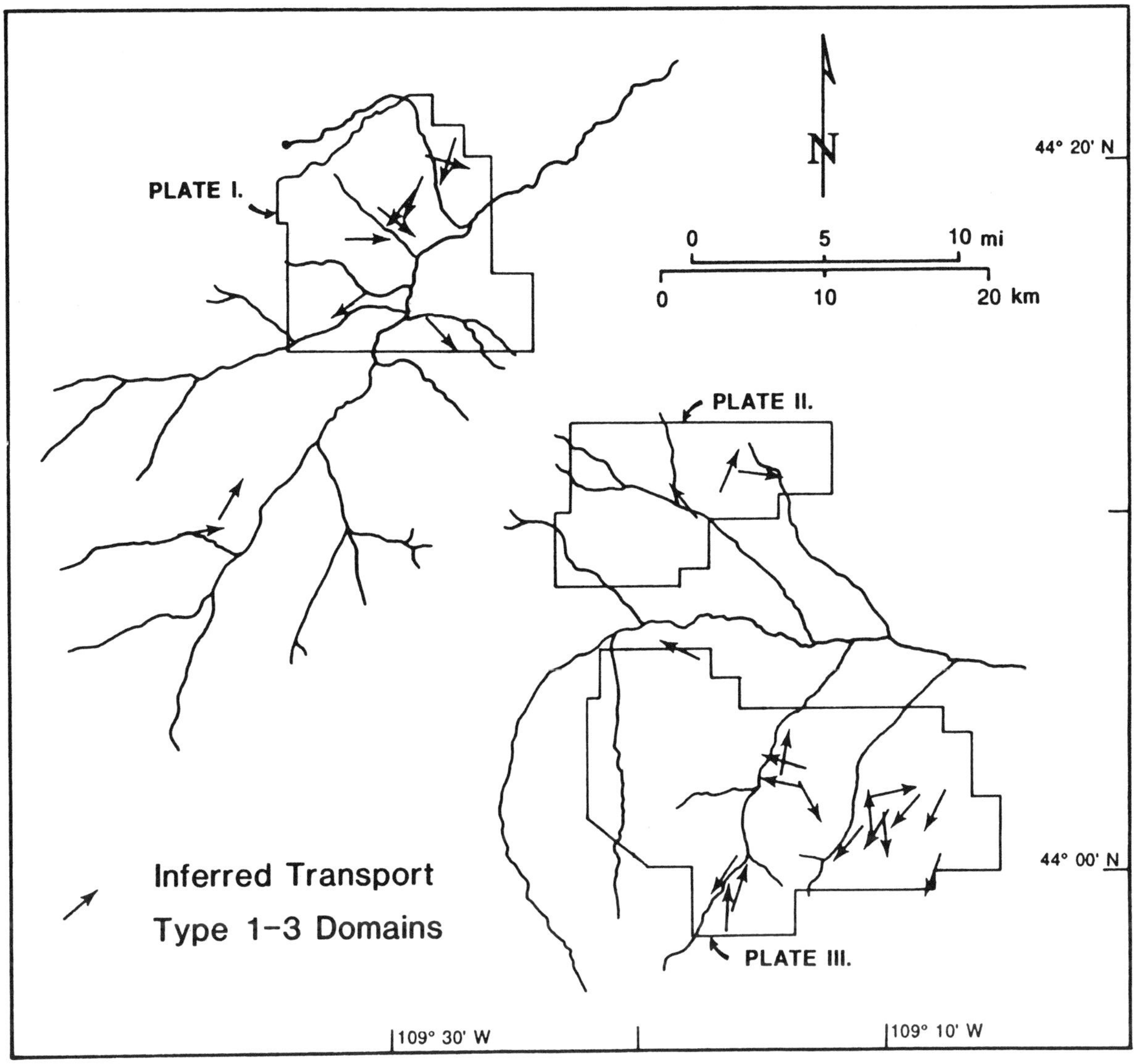

Figure 26. Scattered transport directions (arrows) inferred from structures in domains of Types 1 through 3 recognized in the east-central Absaroka Range. These reflect the simplest interpretations of structural geometry rather than measurements of precise kinematic indicators, which are generally lacking in these rocks because deformation occurred prior to widespread lithification. Streams and rivers are as labeled in Figure 1b.

acceptable because of relations demonstrating subsurface deformation; furthermore, this deformation occurred after, rather than during deposition, although there is evidence that some of these domains contained initially unsorted, unbedded debris-flow deposits.

Of the limited previous work on Type 4 structural development, the descriptive observations of Hay (1954) are the most consistent with those of this study, although the most thoroughly developed of his multiple working hypotheses, volcanic interstratal injection, fails on several counts, as he admitted (1954, p. 617–618). While this hypothesis does account for the "intrusive" contacts between chaotic intervals and surrounding strata, several lines of evidence argue against an origin by primary injection: (1) absence of syndeformational volcanic material in the majority of cases, (2) lack of volcanogenic alteration (Hay, 1954, p. 614–615), and (3) presence of gradational contacts between chaotic "tuff-breccia" and well-stratified epiclastic strata (Hay, 1954, p. 617). Ultimately rejecting this hypothesis, Hay concluded with a minimum of interpretation by summarizing his observations in a manner nearly entirely consistent with the present interpretation of seismic liquefaction, which was virtually unknown at the time of his writing: "In conclusion, the unusual structural relationships of tuff-breccia can be best explained by post-depositional movement of several sheets of tuff-breccia deposited at the surface and buried by detrital sediments. No satisfactory explanation can yet be given for the cause of this underground movement, which was probably forceful." (Hay, 1954, p. 619).

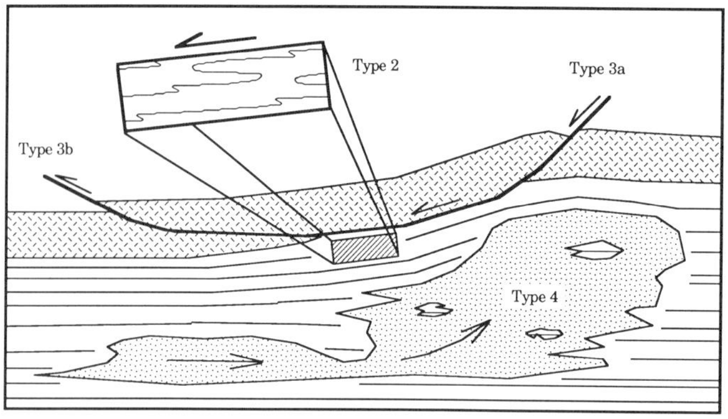

Figure 27. Diagram showing hypothetical relation of liquefaction and dismemberment in a Type 4 domain to secondary deformation in structurally higher domains of other types. Gravitational instability is generated by equilibration of upper strata to deeper liquefaction and flow. For simplicity, no Type 1 folding is depicted here, but it could occur in place of Type 3 faulting in a buckled layer overlying a zone of depth creep. The layer-parallel shearing recorded by Type 2 structures could occur either in response to drag forces in the footwall of gently dipping faults, as shown here, or within a zone of depth creep. This figure is schematic only, is not to scale, and does not attempt to portray all varieties of secondary instabilities caused by liquefaction.

Liquefaction is the only deformational process known to the writer that satisfactorily explains all six of the interpretive characteristics of Type 4 domains outlined above. Furthermore, the close spatial and temporal association of liquefaction domains with those of all other types suggests that the local gravitational instabilities that created structural Types 1 through 3 were precipitated by liquefaction-induced tilting and buoyancy equilibration, as summarized in the hypothetical example of Figure 27. In other words, the numerous Type 1, 2, and 3 structures simply record different responses to local instabilities caused by liquefaction. Hence, the predominant Type 4 deformational process—liquefaction by seismically induced generation of pore fluid overpressures—is the common thread throughout the unusual middle Eocene deformation in this part of the Absaroka Range.

Discussion of classification scheme

As with many attempts to impose order onto nature, the domain classification scheme devised here has its limitations and idiosyncrasies, which must be identified in order to clarify its virtues. What conclusions regarding deformational processes can justifiably be drawn from the distribution of domains classified by this scheme?

One fundamental limitation has already been established: it makes only second-order genetic distinctions between structures most probably caused by a common first-order genetic process, i.e., various responses to pore fluid overpressures. In addition, the scheme's value is arguably diminished in that structural styles within domains are commonly mixed, and the boundaries between domains are typically unmappable. Furthermore, the distribution of deformational-style domains is controlled in part by haphazard variations of local stratigraphy, sedimentology, and textural characteristics rather than systematic variations in externally determined properties such as deformational process, magnitude of applied stress, or paleotopographic relief.

Nonetheless, the scheme does provide a necessary descriptive basis for cataloging and distinguishing among a wide variety of structures. Its most useful application is in describing the second-order responses that occurred within individual domains. A coherent synthesis of this regional deformation is devoid of context if it cannot be related to these domains on an individual basis—the way they are encountered in the field.

INFINITE-SLOPE COMPUTER MODELING OF (TYPE 1) FOLD DEVELOPMENT IN THE AYCROSS FORMATION, TIMBER CREEK AREA

Introduction

Each of the structural styles classified by the above four-fold classification scheme formed under specific mechanical conditions in different structural positions. Of all of the fold domains encountered in the study area, the spectacular Type 1 fold train in the Timber Creek drainage south of the Greybull River (Fig. 16a) offered the best opportunity to model the conditions of local deformation. As described above, these folds appear to have developed by gravity-driven movement of the folded sequence over an uninvolved "basement." These characteristics led to deformational modeling via Infinite-Slope Analysis theory, which is commonly used in soil mechanics to evaluate rates of downslope creep and the potential for landsliding. The theory's application to this study was tailored to estimate the downslope displacement (strain) rate, and thus, the duration of structural development in the Type 1 Timber Creek fold train.

The interplay between structural geology and soil mechanics is a natural connection in this case, because the folds under consideration developed very near the land surface in cohesive, but unlithified or only partially lithified sedimentary strata, presumably by gravitational processes. The local middle Eocene stress field that led to the deformation of the materials at the site is believed to have been induced by a combination of local tilting and topographic relief. Post-depositional tilting occurred here and elsewhere in the east-central Absaroka Range, probably as an adjustment to widespread seismic liquefaction in the subsurface. The proposed paleotopographic relief developed during an intra–Aycross/Wapiti episode of erosional downcutting on the margins of the rapidly building Absaroka volcanic-volcaniclastic pile. The stratigraphic position of the series of folds in question places them very near the surface during this middle Eocene erosional episode. Their asymmetric geometry is interpreted as the result of buckling of strata in the near-surface "rigidity" zone overlying a zone of "depth creep" (Ter Stephanian, 1974). The numerical results obtained through this modeling suggest a time span in the range 10^1 to 10^2 yr for the formation of each fold wavelength.

Method and procedure

Modeling was carried out using modified versions of INSLOPE, a BASIC language program provided by Dr. Peter Bosscher, Department of Civil and Environmental Engineering, University of Wisconsin—Madison, which is based on elements of Infinite-Slope Analysis theory developed in large part by Ter-Stepanian (1963, 1974). This theoretical approach has its basis in the observation that end effects can be neglected in predicting deformations within a long slope, just as they would not exist in a hypothetical slope of infinite length.

INSLOPE was modified in several versions to accommodate value ranges for numerous input parameters. After establishing the relations between parameters, the program was further modified to use a range of values of slope angle, β, as an input variable (INSLOPEmodβ), and was constrained by estimates of other soil parameters derived in part from geologic field observations. The program was run in *Microsoft BASIC* on a Macintosh Plus computer, supplemented by *Microsoft Works (database application), Switcher,* and *Cricket Graph* software for data manipulation and analysis.*

Geology

Most of the geologic observations used in this computer modeling were made on the west side of the Timber Creek drainage of the Greybull River west of Meeteetse, Wyoming (Fig. 28). The well-developed folds at this locality occur in volcaniclastic rocks of the Aycross Formation (middle Eocene).

These fold structures are just one type in a broad spectrum of structures formed in the Absaroka Range during middle Eocene time. None of these individual structures extends for distances of more than several hundred meters, although numerous local structures with very different styles, orientations, and directions of mass transport are closely juxtaposed. This complex structural pattern is incompatible with orogenic deformation. Regional detachment faulting is considered unlikely because no detachment surface has been found, and because structures are of extremely variable orientation and intensity. In order for intense non-orogenic folding to occur at such shallow depths, the sediments involved could not have been completely lithified. For these and other reasons detailed more fully in other sections of this volume, it is believed that this suite of structures probably formed in response to local gravitational instabilities induced by extensive liquefaction of unlithified sediments in the subsurface.

Cohesive but unlithified during deformation, the folded strata are a heterogeneous succession of thin- to thick-bedded, interstratified conglomerate, sandstone, siltstone, and mudstone, with these size fractions admixed to form a wide range of textures. Folds are open to tight, with gently to moderately inclined axial planes (Fig. 29a). Wavelengths range from about 30 ft (10 m) in thinner, weaker mudstone beds to more than 150 ft (50 m) in thickly bedded, more rigid volcanic sandstones with clay and silt matrix. On the basis of the fold geometry and probable topographic conditions during deformation, this folding is considered to have occurred by buckling in a "rigidity zone" overlying a zone of "depth creep" (Ter-Stepanian, 1974) with compression induced by differential creep rates at a slope-break or slope-toe (Fig. 29b). Prevailing slope angle ($\beta1$) and vertical thickness of involved slope materials (H) were estimated from field photographs for input into the model. Shortening was calculated by assuming parallel (concentric) folding from estimated bed lengths and fold tightness, and was in turn used to calculate the time necessary for folding.

*Mention of brand names is for purposes of documentation only, and does not constitute a commercial recommendation or endorsement of any kind.

Description of the computer model

The modified version of INSLOPE used in the final model is listed in Appendix II. This illustrates the form of the program used for this geological application, which focused on defining creep rates rather than examining the influence of changes in surcharge (overburden pressure in addition to contribution from slope thickness, in units of pounds per square foot, spf) or height of the ground-water table. For the purposes of this study, additional subroutines originally written to investigate these extraneous factors were omitted from INSLOPEmodβ. The program calculates values of creep rate at the land surface (in in/yr), height (in ft) of the rigidity zone above the zone of creep, factor of safety, and the critical height (in ft) of the ground-water table.

Parameters needed for INSLOPEmodβ that cannot be estimated directly from field observations were arrived at by a combination of geological and soil mechanics considerations. For example, cohesion (C) and angle of internal friction (ϕ) were estimated from knowledge of the composition of the sediments and estimates of surcharge (q assumed 0). The value for the height of the ground-water table ($H_w = H/2$, where H represents the vertical thickness of the involved slope) is an educated guess based on high-rainfall paleoclimatic conditions. The value for the coefficient of flow (L) is constrained only by comparison with values used by previous workers in an application of INSLOPE to gravelly clay till; the error associated with the present use of this value is unknown. However, values for this parameter are extremely difficult to obtain, so this value was used in hopes that it is within at least an order of magnitude. Table 3 summarizes the parameters and their values used in the model.

Discussion

Several variations of INSLOPE were run in order to obtain data sets that would yield an understanding of the relation between some of the input and output parameters. Plots illustrating several of these relations are shown in Figures 30 and 31. After establishing these relations, the final version (INSLOPEmodβ) provided the data used to establish the creep rate, to examine the potential for deformation by creep versus sliding, and to evaluate

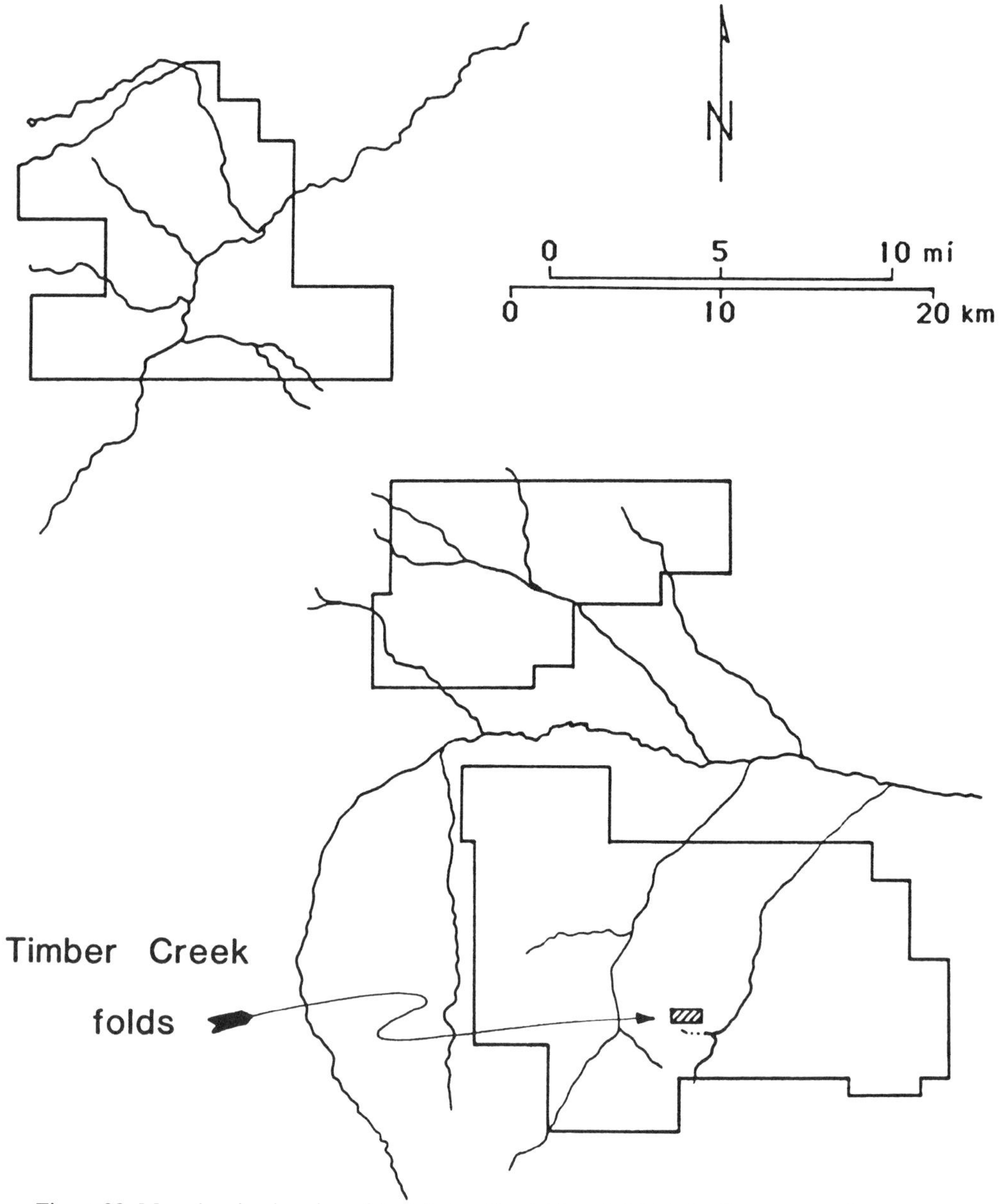

Figure 28. Map showing location of the Timber Creek Type 1 fold train. Streams and rivers are as labeled in Fig. 1b.

the general validity of the model under the specified conditions assumed for deformation. These data are portrayed in the plots of Figures 32 and 33. The following discussion focuses first on relations revealed by plots from the initial data sets, and then considers the data from INSLOPEmodβ.

Factor of safety. Factor of safety (F) is a dimensionless ratio used in soil mechanics and civil engineering to define the likelihood of landsliding failure. For F values greater than unity, a system is considered "safe"; that is, the system is either stable (motionless) or will experience only creep movement. For F values less than unity, sliding is imminent. Figure 30a plots the relation: safety versus slope angle. It is clear that safety decreases logarithmically with increasing slope angle. In Figure 30b, the quasilinear relations safety versus angle of internal friction is seen to vary with surcharge. As expected, safety increases with increasing internal friction. However, the relation between surcharge and safety is not as intuitively obvious; safety is significantly reduced at low surcharge values, but this effect drops off rapidly as surcharge increases. Finally, Figure 30c illustrates the direct linear proportionality of safety versus cohesion, with failure unlikely (F $\sim$1.3), even in cohesionless materials under the conditions assumed for the formation of the Timber Creek folds.

Creep rate. Figure 30a plots the linear relation creep rate versus slope angle, with variations in creep rate due to changes in surcharge. Consistent with the safety trends of Figure 30b, materials tend toward greater mobility with decreasing surcharge; the

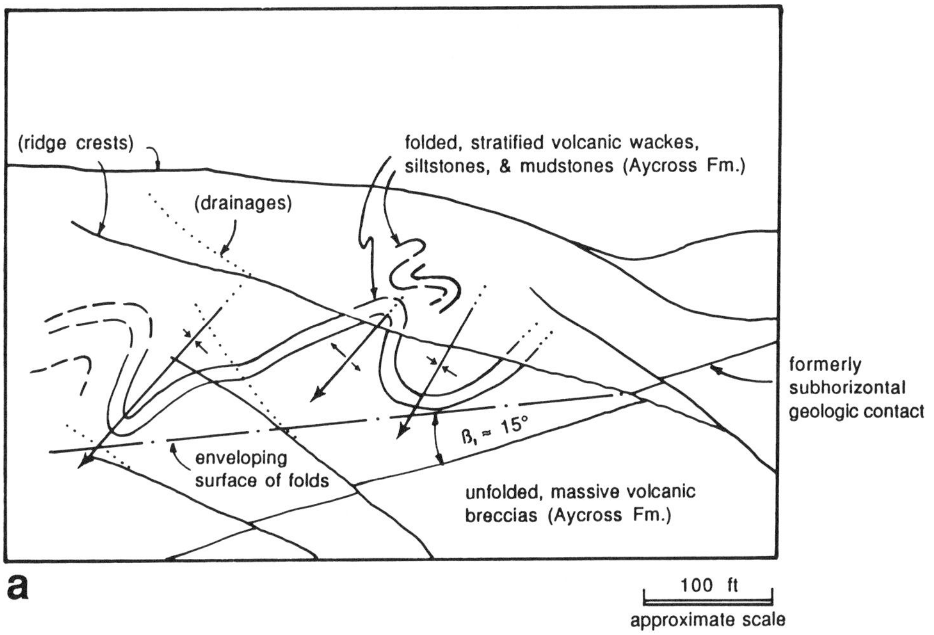

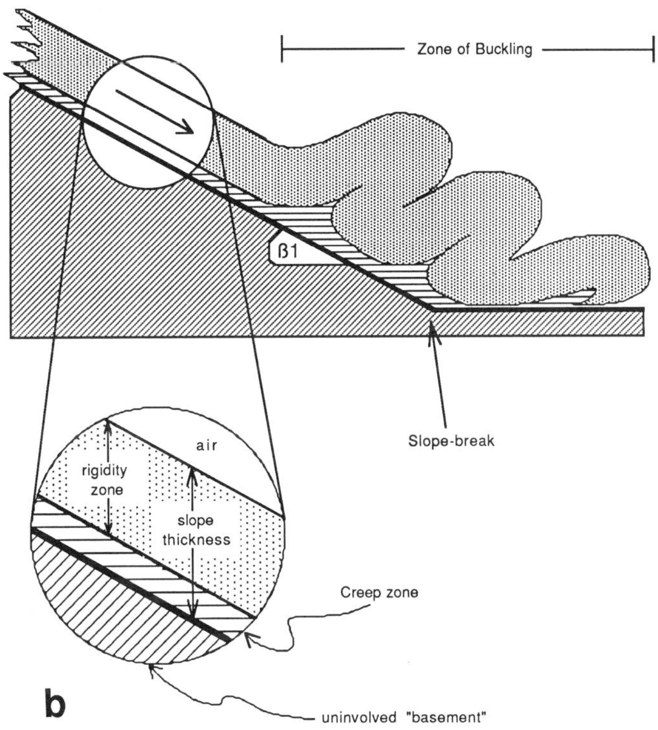

Figure 29. (a) Interpretive photosketch of Type 1 fold train 1 mi west of Timber Creek at ~UTM 12TXD4476. See Figure 16a for photograph of same folds. (b) Schematic representation of fold development in a rigidity zone overlying a zone of depth creep. Buckling is induced by the difference in creep rates above and below slope break.

creep rates of Figure 31a diminish greatly with surcharge values, representing geologic burial. As a guide to interpreting this graph, q = 10,000 psf might result from burial by ~140 ft of gravel, or ~60 ft of sandstone, or ~50 ft of basalt (based on surcharge values of Crozier, 1986, p. 46, and densities of Carmichael, 1984). Figure 31b, also of creep rate versus slope angle, is presented to show that creep rate is essentially independent of cohesion in the range C = 0 to 400 psf. This observation is reassuring, because cohesion is assumed to be zero in the version of the program (INSLOPEmodβ) used to determine the time span of folding. Even if cohesion reached 400 psf, creep rate would decrease by less than 4 in/yr, introducing a maximum error of about 6 percent for slope angles lower than 25°.

The relation creep rate versus angle of internal friction is plotted in Figure 31c. Clearly, any errors made in choosing an appropriate angle of internal friction will not significantly affect creep rate. Once again, however, the point is clearly made that changes in surcharge *do* affect creep rate. Still, even gross errors in surcharge estimates would translate into <25 percent error in creep rate under the conditions specified for this run.

Creep rate versus cohesion in the range C = 0 to 2,000 psf describes a concave-upward exponential curve (Fig. 31d), with creep rates diminishing by as much as 75 percent toward the high end of this range. Based on the inferred surcharge (q = 0) and consolidation history (normally consolidated), however, it is unlikely that cohesion reached values anywhere near this probable maximum prior to deformation. Nonetheless, if in the course of further study these assumptions are shown to be incorrect, this plot would provide limits of creep rate error over the possible range of cohesive strength.

Results of INSLOPEmodβ. Figure 32 is useful in evaluating the potential for failure by landsliding under the conditions specified in Table 3, used in INSLOPEmodβ. Figure 32a (safety versus slope angle) reveals that sliding is only probable at slope angles of ~20°, considerably greater than ≤15° incline suggested by field evidence. Similarly, Figure 32b (safety versus creep rate)

TABLE 3. PARAMETERS AND VALUES USED TO CONSTRAIN INFINITE-SLOPE MODEL OF TYPE 1 FOLD DEVELOPMENT

Parameter, symbol	Estimate Used	Basis of Determination
Slope angle, ß	0-20°	Field photo; for calculations: ß1 ~ 15°, ß2 ≤ 7°
Angle of internal friction, Ø	25°	Composition
Slope thickness, H	100 ft	Field photo
Height of GW table, Hw	50 ft	Paleoclimate, high rainfall
Moist density, Gm	115 pcf*	Composition
Saturated density, Gw	125 pcf	Composition, porosity estimate
Cohesion, C	0 psf	Composition & assuming •drained, •normally consolidated, •q = 0
Pseudo-cohesion (roots), Cr	0 psf	Negligible for H = 100 ft
Surcharge, q	0 psf	Deformed ~ at surface
Seismic coefficient, Ks	0 g	Assumed 0 in calculations for creep rate subroutine
Coefficent of flow,	8.33×10^9 sec^{-1}	Preexisting INSLOPE application
Bond coef. of mobilized shear strength, θ_0	8.33°	θ_0 ~ Ø/3
Shortening:		Field photo †
in percent	30 %	[(Li - Lf)/ Li] x 100
in feet	335 ft	(Li - Lf)

* pcf = pounds per cubic foot

† Li = initial length and Lf = shortened length of measured bed

shows that even at the 28 in/yr creep rate obtained from Figures 31d and 33b for the inferred deformational conditions, safety remains well above the sliding threshold.

The relation between the height of the rigidity zone overlying the zone of creep and creep rate is of interest. The point to be made in Figure 33a is that the rigidity zone becomes thicker at higher creep rates, occupying nearly 85 percent of the total slope thickness at the creep rates inferred for deformation. This can be rationalized on a more intuitive level by envisioning landsliding surfaces as infinitely thin zones of greatly accelerated creep.

Figure 33b is the final plot generated from the results of INSLOPEmodβ, and is the source of the creep rates used to determine the likely duration of folding in the Timber Creek locality. An essential observation is that, under the conditions specified in running INSLOPEmodβ, creep ceases at inclinations of $\leqslant 7°$. Assuming that the slope-break or slope-toe separated an upper segment of $\beta_1 \sim 15°$ from a lower segment of $\beta_2 \leqslant 7°$, the velocity difference between the motionless lower portion and the still-moving upper portion of the 100-ft-thick slope would have been 28 in/yr. Taking the 335 ft (4,020 in) of shortening involved in creating one wavelength of these folds (Table 3), this yields a deformation period of 144 yr per fold wave. A likely minimum for β_1 is 10° (creep rate = 10 in/yr), which would extend this time interval to ~400 yr, and a likely maximum is 18°, which would shorten it to ~100 yr, assuming other variables are adequately constrained.

In view of all of the parameters that can be adequately evaluated, it is considered likely that this deformation period is accurate to within at least an order of magnitude. From a geological perspective, the time periods involved would be described as noncatastrophic, but they are probably much shorter than the time span of folding caused by collisional plate tectonic orogenesis.

Conclusions

Several variations of the INSLOPE program have provided data that reveal the general relations among input parameters (slope angle, angle of internal friction, cohesion, surcharge) and output values (factor of safety, creep rate). It is clear from plots of these data that, over the probable range of deformational conditions, safety and creep rate are more sensitive to slope angle and surcharge than to the angle of internal friction or cohesive strength.

The folds of the Timber Creek Type 1 domain are rootless, asymmetric, and locally restricted structures consistent with gravitational deformation by buckling in a rigidity zone overlying a zone of depth creep. (No evidence is seen at this locality of deformation by layer-parallel shearing within the zone of depth creep, although the sheath folds and other structures of nearby Type 2 domains may have formed by shearing at creep rates in this position.) According to this model of Type 1 deformation, folds would most likely have developed in sequence, with those at the front of the creeping slope developing earlier than those higher on the slope. If given a dominant upper-slope angle of 15° and a gentler lower-slope angle of ≤7°, the model yields a time interval for the folding of each full "wave" of about 145 yr. However, with many of the requisite input variables lacking confidence limits, it may be more legitimate to relax this span to between several decades and several centuries. Because the deformed strata were never exposed at the surface, despite the inference that folding occurred at very shallow structural levels during a period of widespread erosion, the maximum duration of deformation is not likely to surpass this range. The lower limit of the local deformational time span must accommodate shortening by folding rather than faulting of these partially consolidated strata. In this light, creep rates requiring a minimum of several decades seem appropriate.

Regardless of the precise duration of folding at this particular Type 1 domain, both geologic field observations and the numerical results of this modeling favor deformation at fairly rapid creep rates, rather than by either instantaneous catastrophic sliding or million-year orogenic processes. Future work should focus on the importance of gravitational processes operative at these intermediate-length time spans (10^1 to 10^2 yr), which may be too frequently overlooked in studies of structural development.

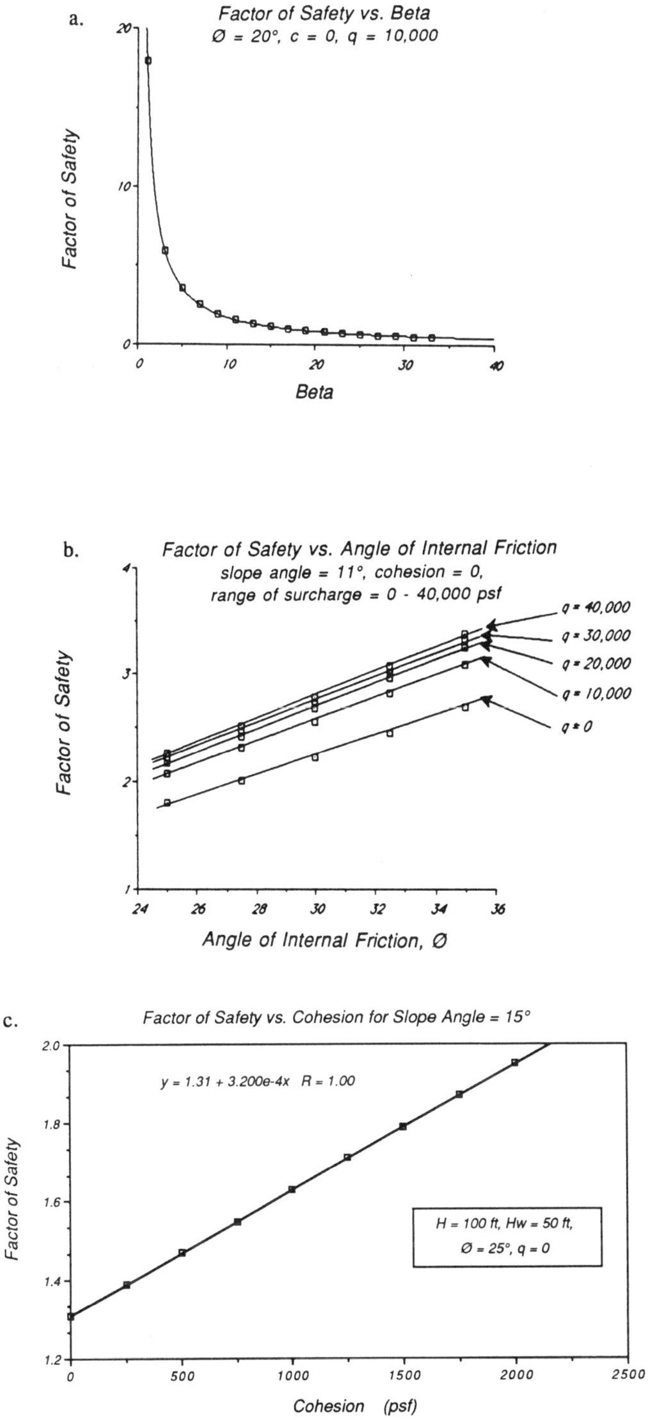

Figure 30. (a) Graph of factor of safety versus slope angle (beta, degrees). (b) Graph of factor of safety versus angle of internal friction (degrees) for a range of surcharge (psf = pounds per square foot) values. See text for discussion. (c) Graph of factor of safety versus cohesion (psf) for a slope angle of 15°. See text for discussion.

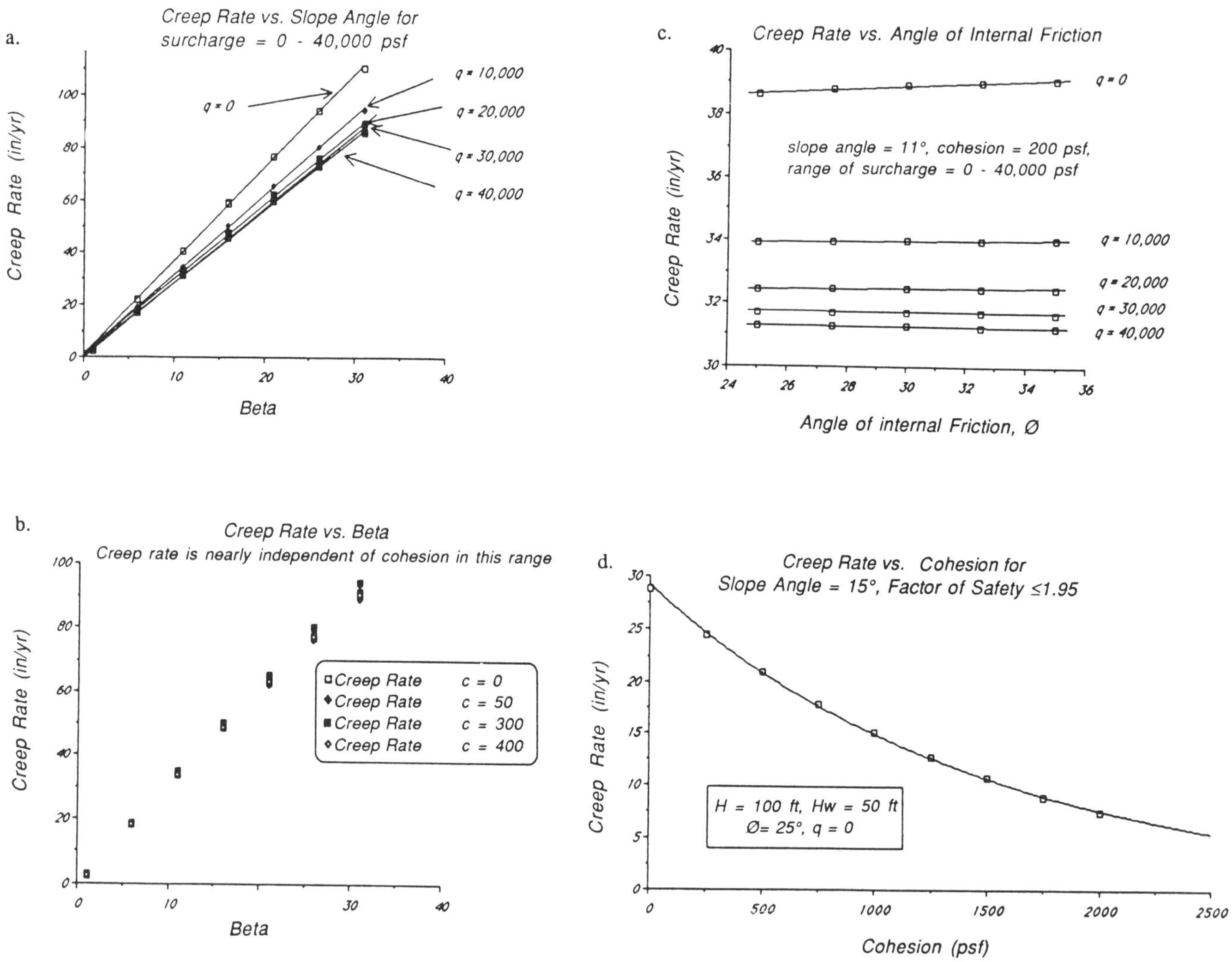

Figure 31. (a) Graph of creep rate (in/yr) versus slope angle (beta, degrees) for a range of surcharge (psf) values. (b) Graph of creep rate (in/yr) versus slope angle (beta, degrees) for a range of cohesion (psf) values. See text for discussion. (c) Graph of creep rate (in/yr) versus angle of internal friction (degrees) for a range of surcharge (psf) values. (d) Graph of creep rate (in/yr) versus cohesion (psf) for a slope angle of 15° and factor of safety $\leq$1.95. See text for discussion.

ROLE OF LIQUEFACTION IN (TYPE 4) DEFORMATION OF THE WAPITI AND AYCROSS FORMATIONS

Introduction

Thick chaotic intervals of dominantly epiclastic and some pyroclastic rocks are widespread in the Aycross and lower Wapiti Formations in the east-central Absaroka Range. Geologic relations at numerous localities indicate that this haphazard structural pattern is due in part to primary depositional characteristics, such as the heterogeneity inherent in many volcaniclastic lithofacies (e.g., laharic debris flows), and the complex interfingering between lithofacies. More significant, however, is the compounding of this primary complexity by secondary, post-depositional in situ disruption, which in turn is largely controlled by local textural characteristics.

Categorized here as Type 4 domains, these intervals of chaotic epiclastic sediments are locally as thick as 550 ft (170 m), but are more commonly about 300 ft (90 m) thick. As massive, turbulent mixtures of disrupted strata and individual sedimentary particles from all size fractions, Type 4 domains contain no discrete fault surfaces, coherent folds, or meso- or microfabrics. Similarly, evidence of significant lateral transport is nearly absent within Type 4 domains. However, many structures from associated domains of Types 1 through 3 do indicate limited horizontal mass transport, and they probably constitute a suite of consanguine responses to widespread deformation in Type 4 domains.

Observations on a wide range of scales support the interpre-

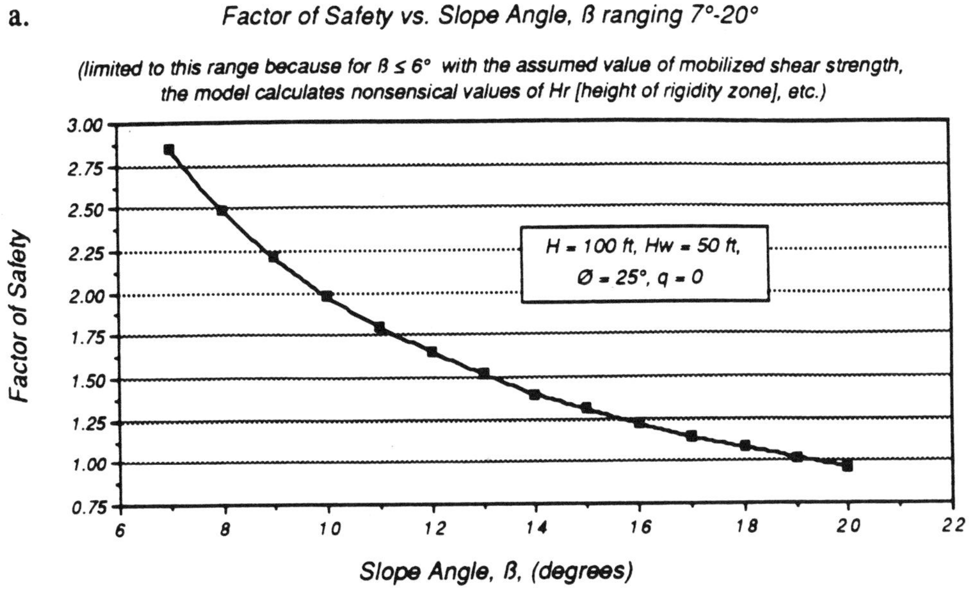

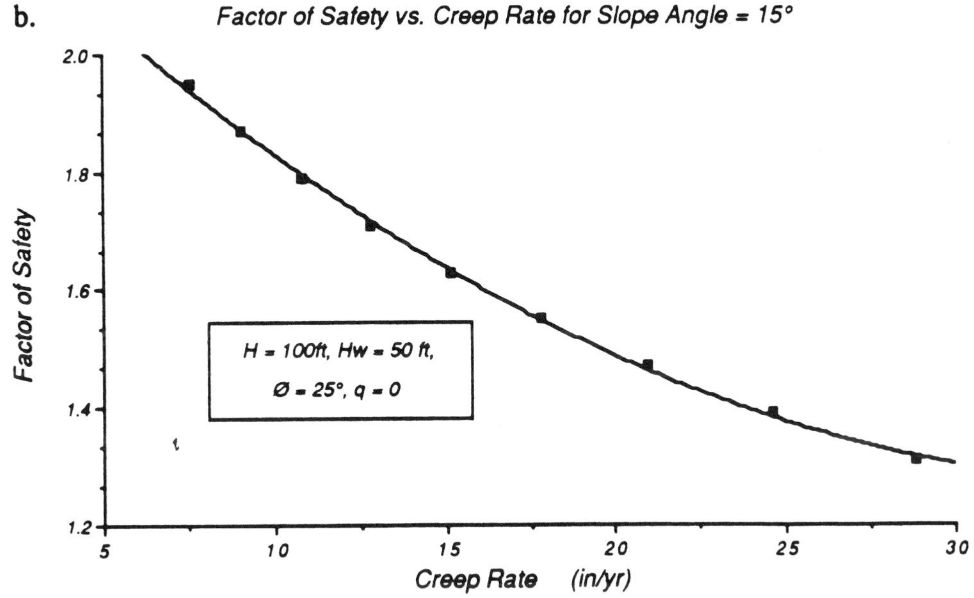

Figure 32. (a) Graph of factor of safety versus slope angle (beta, degrees) over plausible range of slope angles. (b) Graph of factor of safety versus creep rate (in/yr) for a slope angle of 15°. See text for discussion.

tation that Type 4 deformation occurred after deposition but prior to extensive lithification. Disruption was localized at depths typically greater than 300 ft (90 m) below the surface, with grain-boundary sliding and some cataclasis as the principal deformation mechanisms. In many Type 4 domains, the local obliteration of sedimentary layering indicates high cumulative strains, and complex cross-cutting relations between these chaotic, structureless masses and the less disturbed neighboring strata testify to deformation in a "liquidized" state (Owen, 1987), i.e., the sedimentary mass flowed while temporarily transformed into a viscous fluid. Analogous "intraformational disturbances" occur in turbidites of the Triassic Inai Group, Japan (Kamada, 1980), in which hydraulically confined, coarse-grained, pebbly sand was intensely disrupted at depth in intervals as thick as 80 m. Many of those folds, faults, and other contortions bear striking resemblances to structures in the lower Wapiti and Aycross Formations of the study area.

Deformed intervals also occur in the Langford Formation (middle to upper[?] Eocene [Smedes and Prostka, 1972]) near Eagle Peak on the southeastern margin of Yellowstone National Park (H. Smedes, personal communication, 1988), but they are thinner than those in the study area and of more limited lateral

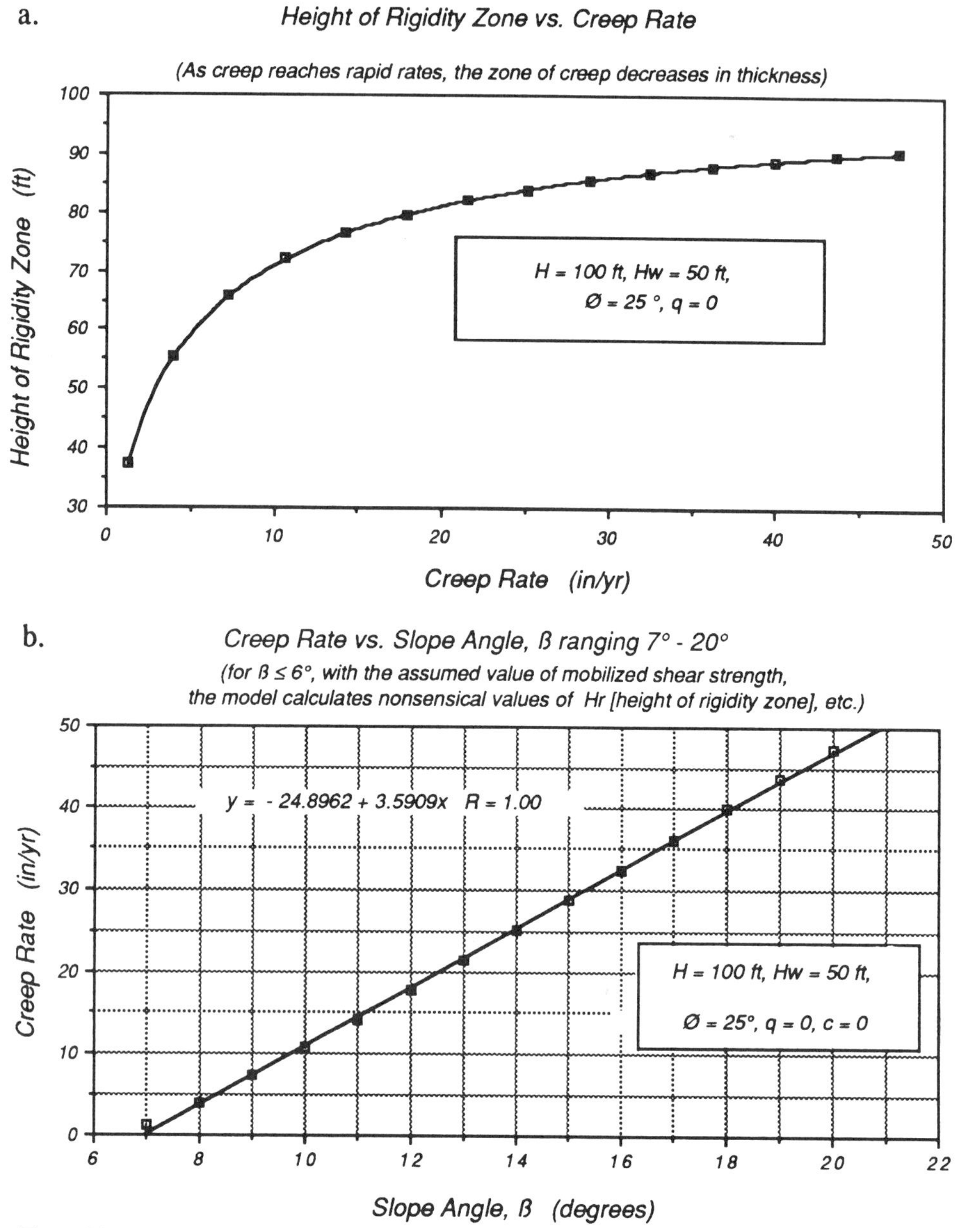

Figure 33. (a) Graph of height of rigidity zone (ft) versus creep rate (in/yr) under conditions inferred for fold development. Thickness of rigid zone increases; hence, thickness of zone of depth creep decreases, as rate of displacement increases. At landslide rates, the "zone of depth creep" diminishes to a discrete slip surface. (b) Graph of creep rate (in/yr) versus slope angle (beta, degrees) under conditions inferred for fold development. Given a dominant upper slope angle ($\beta 1$) of 15°, the model yields a creep above the slope break of 28 in/yr, while below the slope break ($\beta 2/7°$), creep is stopped completely. See text for additional discussion.

extent. These prelithification disturbances are clearly more recent than those in the Aycross and lower Wapiti Formations, and although the structures formed are similar, the triggering mechanism may or may not have been identical. As is discussed below, a wide variety of stimuli can trigger localized liquidization. Nonetheless, their occurrence further illustrates the susceptibility of unconsolidated volcaniclastic sediments to liquidization.

Liquidization: Distinguishing among processes

Liquidization of unlithified sediments may occur by any of four distinct processes, all of which allow the mass to act and deform as a viscous fluid or as a plastic substance of neglible yield strength: (1) thixotropy, (2) sensitivity, (3) fluidization, and (4) liquefaction (Owen, 1987).

Thixotropy and sensitivity. Both thixotropy and sensitivity are common only in high-moisture cohesive clays; they both occur when sudden loading or shaking converts a loose, metastable packing of clay flocs into a dispersion of individual clay platelets in a water base (Owen, 1987). The chief difference between thixotropic and sensitive (quick) clays lies in the fact that strength is restored to thixotropic clays during reflocculation, whereas sensitive clays regain their former strength only if a stable packing is achieved under dry conditions (Owen, 1987).

Fluidization and liquefaction. In contrast to the above processes, fluidization and liquefaction operate in cohesionless materials (or in materials whose cohesion has been reduced to nearly zero), and both achieve a liquid state when interstitial fluids bear the weight of the grains. However, the mechanics of this fluid support differ considerably between the two processes. In a fluidized medium, the upward flux of fluid through an open system exerts a frictional drag force that more than offsets particle weight (Owen, 1987). In this condition, grains are transported upward by the moving fluid. On the other hand, liquefaction results from the reduction of strength that accompanies a buildup of the pore fluid pressure to the value of the confining pressure. As this condition is reached, the effective confining stress approaches zero, particle weight is carried by the pore fluid, interparticle friction is greatly diminished, and the strength (shear resistance) of the medium is lost (Martin and others, 1975; Castro, 1975; Finn and others, 1977; Seed, 1979; Owen, 1987). Of these four processes of liquidization, only liquefaction is likely to have caused the areally widespread subsurface deformations described here from the lower AVS, or those in the Inai Group turbidites (Kamada, 1980). However, other liquidization phenomena may also have occurred locally during liquefaction-related deformation.

Mechanical theory of liquefaction and cyclic mobility

Liquefaction. Two primary types of liquefaction are differentiated by their affinity with either static or dynamic loading conditions. Static loading of the pore fluid is brought about by an increase in pore fluid pressure by an internal stimulus. In regions with geothermal gradients above 15°C/km, confined groundwater is susceptible to static pressurization via "aquathermal pressuring" (Barker, 1972). The implications of static thermal pressurization may well be significant for Eocene deformation in the high-heat-flow regime that undoubtedly accompanied Absaroka volcanism.

Dynamic liquefaction is brought about by external loading and may be either monotonic (impulsive), caused by a single loading event, or cyclic, due to repetitive loading (Castro and Poulos, 1977; Davenport and Ringrose, 1987; Owen, 1987). An impulsive load causes the collapse of the granular structure, compacting the medium to cause a sharp increase in pore fluid pressure. Owen (1987) suggested that rapid sediment deposition is one natural trigger available for the impulsive liquefaction of underlying unconsolidated sands. This hypothesis can be logically extended, particularly in the case of a rapidly building volcanic terrane such as the Absaroka Range during middle Eocene time, to include the emplacement of lava flows and laharic debris flows as likely impulsive triggers for the localized liquefaction of underlying saturated sediments.

An extreme example of impulsive liquefaction warrants brief digression here. Recent experimental study was prompted by the recognition of blast-induced liquefaction associated with weapons testing. Impulses generated by man-made explosions are of short duration, reaching peaks stress levels in about one millisecond (Veyera and Charlie, 1987)—several orders of magnitude faster than the impulsive loading associated with geologic deposition. Nonetheless, true explosions do occur in nature, as evidenced by shocked mineral grains in meteorite impact structures and in calc-alkaline volcanic arcs (Rice, 1985, 1986a, b; Carter and others, 1986; Loper and McCartney, 1988). For many years, geologists have speculated that other unusual deformational phenomena may be linked to geologic explosions of mammoth proportions, such as the "overthrusts" associated with the Ries structure in Germany, and the Heart Mountain fault in the Absaroka Range (e.g., Bucher, 1933).

In the strictest sense, a shock wave is a compressional front that travels through a medium at supersonic speeds (American Geological Institute, 1980). Seismic (acoustic) energy thus cannot transmit true shock waves, and volcanic eruptions causing actual shock waves probably are not common geologic events. Shock-induced impulsive liquefaction may occur in nature, but it is no doubt uncommon in the geologic record.

Returning to the realm of more likely scenarios, areally widespread liquefaction is typically the result of cyclic rather than impulsive loading, and it is well documented in association with earthquakes (e.g., Fuller, 1912; Hansen, 1965; Seed and Wilson, 1967; Kawasumi, 1968; Ishihara and Koga, 1981; Tohno and Yasuda, 1981; Wang, 1981). Seismic oscillations passing through unconsolidated sediment induce cyclic shearing stresses that destroy the weight-bearing grain-to-grain contacts in progressively increasing numbers, gradually transferring this weight to the pore fluid (Seed and Idriss, 1982). If the permeability of the sediment is insufficient to allow escape of pore fluid at the same rate as it assumes the load from the collapsing granular structure, the fluid pressure rises. As seismic shaking continues, so does compaction and hence fluid pressurization, eventually leading to a condition in which the pore fluid pressure equals the confining stress, and strength is lost. At this point, the mass flows readily, developing high shear strains in the presence of driving shear stresses. This sequence of events is illustrated in the simultaneous experimental response curves of Figure 34. The importance of this process is summarized by Seed and Idriss (1982, p. 69):

> Liquefaction of a sand in this way may develop in any zone of a deposit where the necessary combination of in-situ conditions and vibratory deformations may occur. Such a zone may be at the surface or at some depth below the ground surface, depending only on the state of the sand and the induced motions.

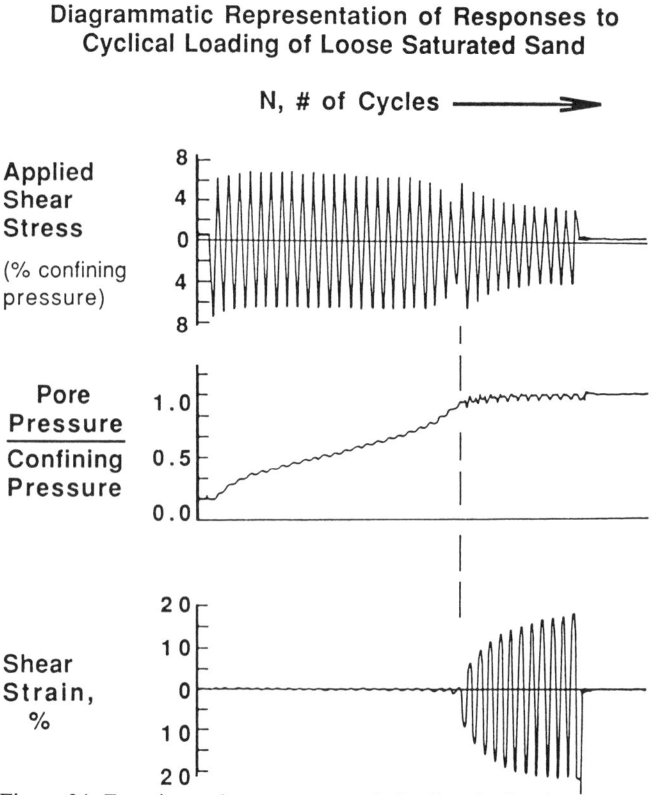

Figure 34. Experimental response records for liquefaction induced by cyclic shearing stresses. Horizontal axis is number of cycles (and time). Dashed vertical line marks number of cycles (time) at which pore pressure becomes critical and shear strain accumulation begins. Modified from Seed and Idriss (1982).

Cyclic loading provides a mechanism for developing large cumulative strains in sediments that are not susceptible to liquefaction under monotonic loads. For example, during impulsive uniaxial loading, dense sands may actually dilate by the opening of microcracks, which decreases the pore fluid pressure and strengthens the medium. The same dense sands may contract under cyclic loads, which pressurizes the pore fluid and weakens the mass (Martin and others, 1975). In both impulsive and cyclic liquefaction, overpressures are sustained only temporarily, with both the amount of increase and the rate of dissipation governed by the medium's permeability (Wong and others, 1975; Oka and others, 1981; Ishihara, 1981, Owen, 1987).

Cyclic mobility. In many coarse-grained sediments, permeability is too high to sustain fluid pressures equal to confining pressures, and the medium cannot achieve truly fluid behavior. Similarly, if sediment density is above a critical threshold prior to the onset of cyclic loading, the overall compaction caused by shaking is insufficient to pressurize the pore fluid above the value of the confining pressure, and it retains much of the character of a solid. Even so, many sediments that cannot truly liquefy because of high permeability and/or density may be susceptible to less severe deformations under the conditions of reduced strength associated with the related process of cyclic mobility. Castro (1975) summarized cyclic mobility induced in triaxial tests using cyclical normal stresses by alternating compression and extension of equal magnitudes. Additional verification of liquefaction and cyclic mobility comes from experiments using cyclic shearing stresses, which more closely imitate the loading experienced by sediments in response to seismic oscillations (Seed and Idriss, 1982). The basic mechanism of initial strength reduction in cyclic mobility is the same as for cyclic liquefaction: pore fluid pressure increases gradually due to small, cyclical, compactional strains—<1 percent per cycle (Castro, 1975)—brought about when a cyclic stress destroys grain contacts and causes the sediment to become more closely packed.

Cyclic mobility differs from liquefaction in that, after the initial pore fluid pressure buildup, the pore pressure equals the confining pressure only momentarily during each loading cycle, and only at this moment does the effective stress become zero. This occurs only when the applied shear stress equals zero, midway between its maximum and minimum values (Castro, 1975; Seed and Idriss, 1982), because the sediment mass expands as microcracks dilate in response to a nonzero stress. As shown in Figure 35, the wave-form of the applied stress crosses the horizontal axis twice during each full cycle, once in dropping from the peak to the trough, and again in rising from the trough to the next peak. The drop in pore pressure at peaks and troughs is attributed to transient dilation (microcrack opening) of the mass. Thus, only for an instant—twice during each cycle—the fluid pressure is maximized because the sediment contains no microcracks, the effective stress is reduced to zero, strength is lost, and an additional increment of strain is free to develop.

An important feature of cyclic mobility is that the cyclic strain (amount of strain per cycle) increases after the initial development of transient high fluid pressures (Castro, 1975; Seed and Idriss, 1982). This effect eventually tapers off, and at some level of cyclic strain, the deforming medium can sustain an unlimited number of loading cycles without further increase in cyclic strain (Seed and Idriss, 1982). The curve of cumulative strain versus time (number of cycles) for deformation by cyclic mobility is thus concave upward, with strain increasing at increasing rates early on, and then increasing at a constant rate to a total strain limited only by duration of the shaking.

Controls on deformational processes. Table 4 lists the factors that, according to Seed and Idriss (1982), most significantly affect the deformation of unconsolidated materials by seismic liquefaction. These elements, many of which interconnect or overlap in reality, can be grouped into three categories: (1) physical properties of the granular medium, (2) environmental conditions, and (3) earthquake characteristics. As engineers, Seed and Idriss (1982) defined these variables in terms of "soil," and some substitutions have been made to fit their scheme to the context of sediment deformation.

As discussed above, sediment density and permeability are two of the critical physical properties bearing on susceptibility to deformation by fluid overpressures. Density must be sufficiently low prior to cyclic loading that the medium will decrease in

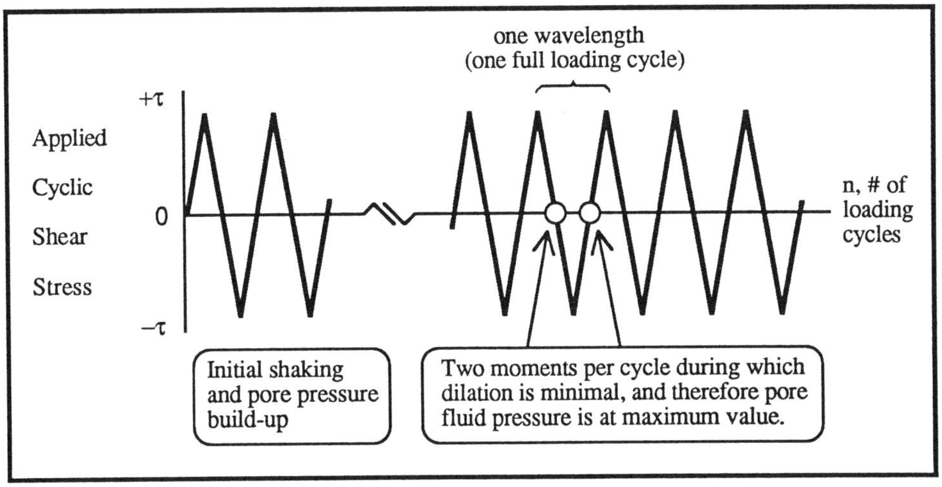

Figure 35. Diagrammatic interpretation of transient pore pressure maximization associated with cyclic mobility after initial pore pressure build-up. When applied cyclic stress is zero, dilation (dilatancy) of mass is minimized, pore fluid pressure is maximized, and sediment is momentarily mobilized. Dilation increases as applied load rises, which decreases pore pressure and restores strength until load again becomes zero.

volume to pressurize the pore fluid (Castro, 1987), and permeability must be low enough to preclude rapid dissipation of pore pressures. In addition, cohesion must be either zero or very low in order to permit the breakage of grain contacts essential to the compaction of the granular structure. Grain characteristics (sphericity, roundness) combined with characteristics of the aggregate (sorting, structure) together influence the angle of internal friction. Because the angle of internal friction and the cohesion jointly determine the sediment's resistance to shear deformation, they are reflected in Table 4 in the sediment's shear modulus. Accurate modeling of pore pressure increases in the heterogeneous and anisotropic sedimentary sequences flanking volcanic terranes would be extremely difficult because many sediment properties (e.g., permeability, density, shear resistance) vary greatly from place to place, and some (e.g., permeability) differ with direction (horizontal versus vertical).

Whereas the loss of strength inherent in seismic liquefaction is brought about by an external stimulus, the deformational driving force typically comes from within the system. Gravitational forces are ultimately responsible for sediment movement, whether instability is manifest in the form of uneven vertical loads (Owens, 1987; Castro, 1987), density inversions, or laterally unconfined slopes.

A distinctive feature of deformation in the liquefied (or cyclically mobile) state is its potential independence from topographically induced instability. In fact, the generation of high pore fluid pressures is more likely on level ground than on unstable slopes, which in most cases, are well drained. However, if a layer is liquefied (and/or liquidized by disturbance of thixotropic or sensitive clays) in the vicinity of a free face, conditions will favor horizontal mass movement of the overlying material. In a few cases, this kind of landsliding has been well documented in the literature (Hansen, 1965; Seed and Wilson, 1967; Updike and others, 1988), and it can be inferred from descriptions of earthquake damage in other writings (Fuller, 1912, p. 48, 61; Nagumo, 1968). Table 4 recognizes the role of lateral confinement in terms of the "lateral earth pressure coefficient."

As an environmental factor, the depth of the water table is of interest for two reasons. First, liquefaction is impossible in the unsaturated zone. Second, the weight of the water column contributes to the vertical confining pressure, which must be offset by pore fluid pressure to cause liquefaction.

TABLE 4. FACTORS AFFECTING SEISMICALLY INDUCED LIQUEFACTION*

Sediment Properties
Dynamic shear modulus
Damping characteristics
Unit weight
Grain characteristics (roundness, sphericity, sorting)
Actual density relative to maximum possible density
Sediment structure (influence on permeability)
Environmental Factors
Depositional processes of sediment
Seismic history experienced by sediment
Diagenetic history experienced by sediment
Lateral earth pressure coefficient
Depth of water table
Effective confining pressure
Earthquake Characteristics
Intensity of ground shaking
Duration of ground shaking

* Modified after Seed and Idriss (1982, p. 70-71), who defined these factors in terms of "soil" rather than "sediment."

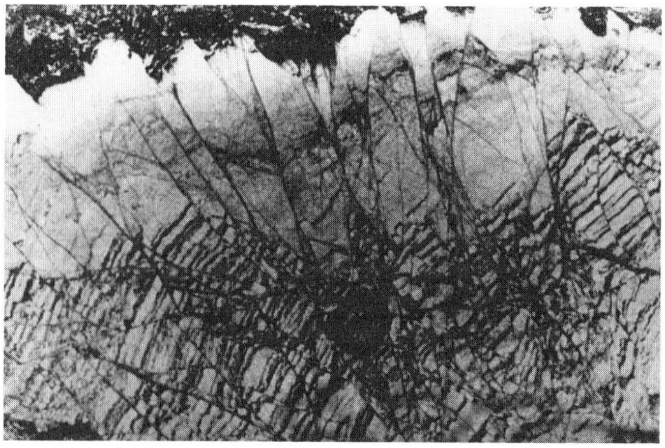

Figure 36. Transposition of lamination by rotation on discrete soft-sediment microfaults in laminated dust tuff in Aycross Formation. Photo in plain light of complete thin section of specimen UW1837-7, collected from folds of Figure 16b on north end of Deer Mountain at UTM 12TXD422783. Field of view 4 cm in long dimension.

Table 4 considers the depositional process and diagenetic history of unlithified sediment as environmental factors. Certainly these elements are determined by the environment, but ultimately they are relevant to deformation only in their effects on the sediment properties discussed above. Similarly, the local seismic history exerts its control on the response to renewed shaking in the form of structural and mechanical changes within the sediment that have been considered in terms of sediment properties. Active volcanic terranes are inescapably subject to frequent small earthquakes; these tremors may prevent the development of strong grain-to-grain contacts.

A potentially important environmental factor, which is missing from Table 4, is the presence or absence of a hydraulic confining layer. Clearly, ground water is more efficiently pressurized in a closed system than in a "leaky" system. Hydraulic confinement may be the only way to achieve liquefaction in high-permeability deposits, such as very coarse sands and gravelly sands. Wong and others (1975) investigated the cyclic loading liquefaction potential of coarse materials, and concluded (p. 582) that "damage due to liquefaction of very coarse sands and gravelly soils in which drainage is not impeded significantly by the presence of fine-grained deposits is likely to be extremely rare even in highly seismic areas." The strength expected of coarse-grained deposits under drained conditions is supported by field observation of the effects of modern earthquakes (Ross and others, 1969).

The final category of influences on seismic liquefaction and cyclic mobility considers the characteristics of the seismic event(s). As expected, the greater the intensity of the seismic oscillations, the more readily the sediment's granular structure is collapsed to pressurize the pore fluid (Castro, 1987). Likewise, the longer this shaking continues, the greater the total compaction that can be achieved, and the more likely it becomes that pore fluid pressures will exceed the confining pressure, allowing liquefaction to develop.

Structural evidence of liquefaction in middle Eocene deformation

A suite of structures within the lower Wapiti and Aycross Formations in the east-central Absaroka Range (Decker, 1988) is interpreted as evidence of prelithification deformation caused by seismically induced pore fluid overpressures. The stratigraphic and geographic distribution of structures in this liquefaction-related suite are summarized in Table 5. These structures developed on a wide range of scales, and reflect deformation in a spectrum of physical states between essentially coherent granular solids and true liquids.

Disrupted bedding. Mildly to thoroughly disrupted bedding is by far the most common indication of prelithification deformation in the east-central Absaroka Range. This disruption occurred by both continuous ("viscous flow") and discontinuous ("brittle failure") deformational processes. Continuous deformation, involving distributed, commonly heterogeneous shearing, is manifest by several styles of folds and the partial or complete homogenization of liquefied strata. In contrast, discontinuous deformation is characterized by displacements on well-defined slip surfaces (Fig. 36; specimen UW1837-4).

Continuous deformation is generally penetrative on the scale of the individual constituent grains except in aggregates of clay-size particles. Granular media deform by penetrative intergranular slip when they become fully liquefied, and if high shear strains are accompanied by locally turbulent particle trajectories, stratification may be entirely destroyed. This is the case throughout large parts of most Type 4 domains, in which massive, structureless clastic intervals are associated with other liquefaction-induced features such as irregular inclusions of relict bedded material. For example, the homogeneity of the Twtb unit across thicknesses of tens of meters is believed to have been caused by mixing in a liquefied and/or cyclically mobile state. Even though this lenticular, discontinuous unit probably was deposited originally as a succession of laharic debris flows, its chaotic internal structure and its anomalous cross-cutting relations to enclosing units strongly suggest that it was disrupted by a post-depositional episode of fluid behavior. This liquefaction is interpreted to have erased most of the original contacts between debris-flow units and other beds within the reactivated slurry. External contacts, both lateral and vertical, between Type 4 domains homogenized in this way and the neighboring stratified units are typically highly irregular; many show intrusive relations or intricate interleaving (Fig. 38).

Disharmonic contortions are preserved in extremely fine detail in laminated claystones (water-laid dust tuffs?), but as seen in the electron micrographs of Figure 37 (specimen UW1837-14), the deformation, though apparently continuous on the mesoscopic scale, is actually nonpenetrative on the ultramicroscopic scale. Clay particles reveal little preferred orientation, and fold

TABLE 5. DISTRIBUTION OF STRUCTURES RELATED TO LIQUEFACTION AND/ OR CYCLIC MOBILITY OF UNLITHIFIED SEDIMENTS OF LOWER WAPITI AND AYCROSS FORMATIONS, EAST-CENTRAL ABSAROKA RANGE

Locality	~UTM	Structure(s)	Map unit(s); rock type†	Interpretation§
Hardpan Cr. to Houlihan Cr.	12TXE2007 to 2210	complex relationships:	Twtb, Twls;	
		•liquefied lenses of Twtb, locally invading overlying Twls beds	ss, cgl, brxia > mudst, sltst	L
		•truncation surfaces: ·steeply dipping: reverse slip normal slip ·gently dipping surfaces thrust slip indeterminate slip	within Twls and between Twls and Twtb	Cover response to L, CM at depth
		•sheath fold(s?)	Twtb; ss, cgl, brxia	L, CM?
		•single-bed inclusions: ·irregular, deformed pods	Twtb; sltst, ss, cgl	L, CM
		•multiple-bed inclusions: ·slighty folded relict beds	Twtb; ss, cgl, brxia	Response to L, CM of adjacent sed.
NE of Corral Cr. to 2503	12TXE2302	liquefaction interval:	Twtb, Twls;	
		•single-bed inclusions: ·irregular, deformed pods	sltst, ss, cgl	L, CM?
		•planar clastic dikes, veins	ss	L
		•soft-sediment microfaults		CM
Cabin Cr. to Deer Cr.	12TXD0887 to 1090	liquefaction interval:	Twls;	
		•irregular upper & lower contacts	ss, cgl, brxia	L
		•multiple-bed inclusions ·slightly folded relict beds ·isoclinal recumbent folds		Response to L, CM of adjacent sed.
		•single-bed inclusions ·irregular, deformed pods ·isoclinal recumbent folds		L, CM?
Rose Butte	12TXD4093	liquefaction diapir within larger anticline-syncline pair	Twtb into Twca; ss, cgl, brxia > mudst, sltst	L
Rose Cr. to Little Rose Cr.	12TXD3992 to 4093	chaotically dismembered interval:	Twtb, Tw;	
		•NE verging asymmetric folds	ss, cgl, brxia	CM?
		•single-bed inclusions: ·irregular, deformed pods	sltst, ss, cgl	L, CM
		•multiple-bed inclusions: ·marginally deformed pods	dense ss, coarse cgl	L, CM
		·rigid, undeformed blocks	mudst, sltst, ss, cgl, brxia	N/A
		•faults within multi-bed inclusions	mudst, sltst, ss, cgl, brxia	CM
		•non-planar clastic dikes, veins	sltst, ss	L
Pickett Cr. to Piney Cr.	12TXD3289 to 3489	chaotically dismembered interval:	Tw, Twf/Twfb;	
		•single-bed inclusions: Tw; ·irregular, deformed pods	sltst, ss, cgl	L, CM
		·isoclinal recumbent folds	ss, cgl, brxia	CM?
		•multiple-bed inclusions: ·marginally deformed pods	dense ss, coarse cgl	L, CM
		·rigid, undeformed blocks	mudst, sltst, ss, cgl, brxia	N/A
		•interbed detachment surfaces		CM?
		•downdropped lava sequence	Twf/Twfb into Tw	N/A
Jack Cr. to Deer Mtn.	12TXD3275 to 4278	widespread deformed interval: •complex truncation sets •tight asymmetric folds •irregular, deformed pods	Ta, Twtb?; mudst, sltst, ss, cgl, brxia, tuff laminated dust tuff sltst, ss, cgl	L, CM
Francs Fk.	12TXD4173	chaotic interval: •irregular, deformed pods •tight asymmetric folds	Twtb, Ta; sltst, ss, cgl Ta; sltst, ss, cgl, brxia	L, CM?
Gwinn Fk. of Dick Cr.	12TXD465748 to 4875	chaotic interval: •sheath folds •thrust faults •multiple-bed inclusions •irregular, deformed pods	Ta; mudst, sltst, ss, cgl, brxia, tuff	L, CM?

† Map unit abbreviations as explained on Plate V. Abbreviations for rock types: mudst = mudstone, sltst = siltstone, ss = volcanic sandstone and wacke, cgl = volcanic conglomerate.
§ L = liquefaction, CM = cyclic mobility.

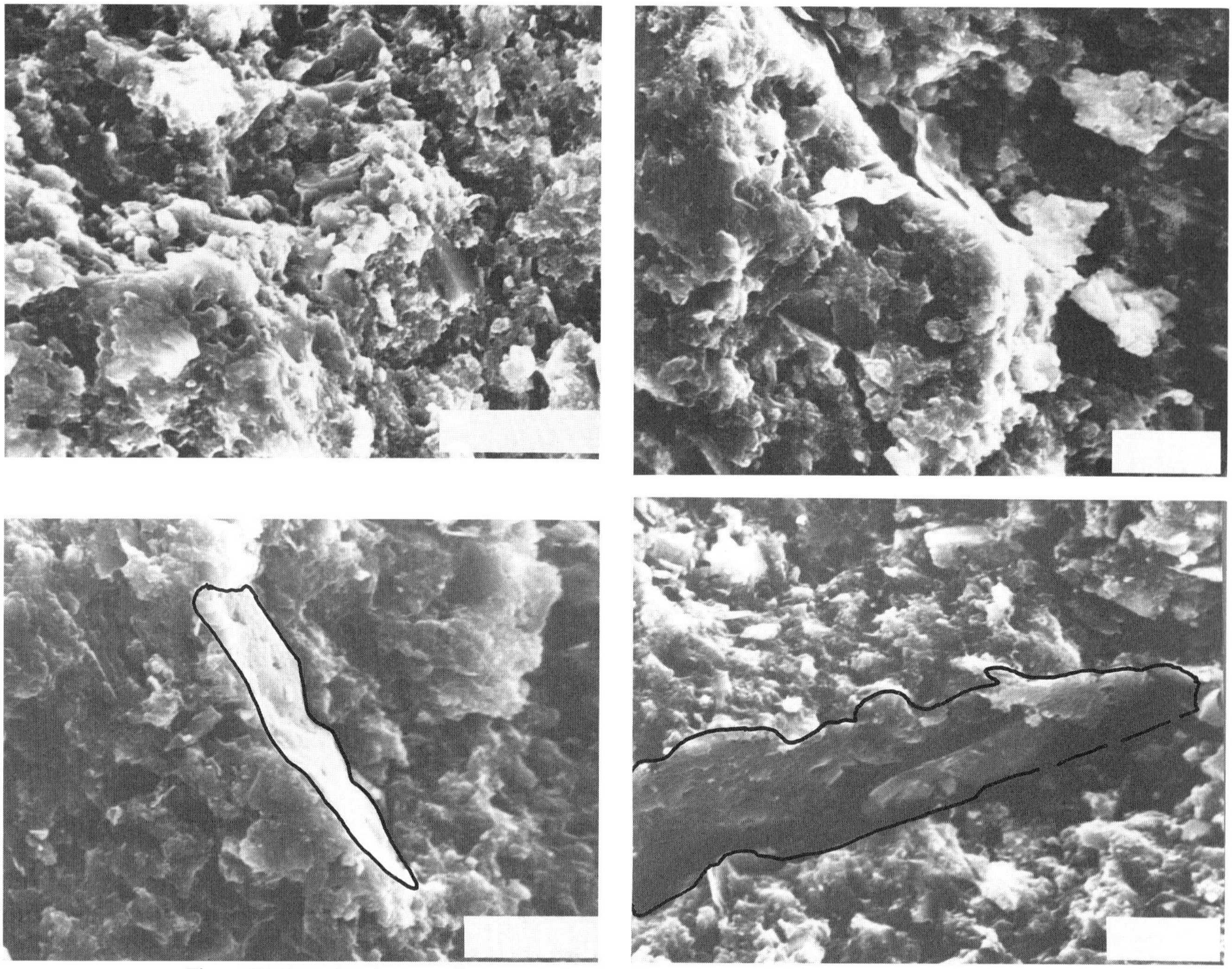

Figure 37. Scanning electron micrographs of textures in contorted laminated dust tuff from Aycross Formation. Clay particles lack pronounced preferred orientation (upper pair of photos). Microfaults in lower pair of photos (surfaces outlined in ink) offset clay particles, demonstrating that much of the folding strain was actually accommodated primarily by discontinuous slip rather than continuous flow. Specimen UW1837-14. Scale bar 10 μm in all photos.

"flow" is accommodated by slip on myriad closely spaced microfaults rather than by penetrative movement of all clay particles.

Stratal disruption by discontinuous deformation occurs on many scales. Varieties include: (1) centimeter-scale offsets in finely laminated sediments; (2) probable meter-scale offsets of splitting surfaces (bedding? incipient foliation?) along gently dipping, locally arcuate truncation surfaces; and (3) fault displacements as large as hundreds of meters. Discontinuous structures at the upper end of this size continuum are generally considered separately as Type 3 domains. Similarly, the largest folds are relegated for descriptive purposes to Type 1 and Type 2 domains, despite the fact that they, too, are probably responses to instabilities created by liquefaction.

Clastic dikes and veins. Two varieties of clastic fissure fillings are recognized as indicating the upward expulsion of overpressurized pore fluids and liquefied sediments. The simplest are those with essentially planar geometries. The rare good examples of these planar clastic dikes are $\leqslant$6 cm wide, and are filled with medium- to coarse-grained sand (specimen UW1837-15). The examples shown in Figure 39 occur at the top of the Twtb unit breccias exposed near Corral Creek in the South Fork Shoshone Valley.

Similar features, though narrower ($\leqslant$1.5 cm) and slightly less uniform, occur in massive volcanic wackes at the mouth of Deer Creek Canyon (UTM 12TXD103904). Here, much of the unliquefied country rock is heavily spotted, probably by thermal-chemical alteration, which accompanied the cooling of several dacitic to dioritic sills and dikes in the area. In the coarse epiclastic breccias superjacent to the quartz diorite Deer Creek laccolith (Rouse, 1933), the clay, silt, and sand matrix is spotted, and the outer 3 to 8 cm of most of the clasts appears bleached and is the most densely spotted, whereas many of the clasts' cores remain

Figure 38. "Feathered" interleaving of liquefied sediments (darker, at left) and relatively undeformed epiclastic beds (lighter, at right) in vertical exposure of Twls unit about 1 mi (1.6 km) SSW of Deer Creek at UTM12TXD0989. The boundaries of liquefaction domains commonly exhibit this sort of irregularity because some beds or parts of beds are more susceptible to liquefaction than others. Field notebook for scale.

Figure 39. Set of three parallel, nearly planar clastic dikes (arrows) injected upward into massive pebbly sandstone at top ot Twtb unit. The overlying sandstone beds of Twls unit remained unliquefied, but formed load casts (largely obscured by shadow at top) projecting into the Twtb unit. The discontinuous Twtb unit as mapped in this study is interpreted in all exposures to have been disrupted by liquefaction and/or cyclic mobility. View to east on Shoshone National Forest Boundary line northeast of Corral Creek at UTM 12TXE251028. Geological hammer for scale.

nearly pristine. The mineralogy of these dark spots is unknown, but alteration is preferentially concentrated in the clasts' highly permeable weathered rind and the somewhat less permeable matrix, compared to the dense unweathered cores. The fillings of clastic dikes in these rocks are unspotted, and are cleaner with respect to muddy matrix materials than their host rocks (specimen UW1837-16). These differences suggest that either the alteration pre-dates clastic dike formation, or alteration occurred after the clastic dikes formed but the sediment fillings remained unaltered. In the latter case, it may be that, during their injection, the sediment fillings were washed clean of clay minerals and/or organic materials, which may have contributed to the formation of spots in the host rock.

Planar and subplanar fissures similar to those filled with the clastic dikes described here have been documented as cutting the land surface in numerous accounts of historic earthquakes (Fuller, 1912; Coulter and Migliaccio, 1966; Kawasumi, 1968; and others). On the basis of eyewitness reports, these fissures are attributed to tensional stresses accompanying swell-like ground undulations caused by seismic surface waves (Fuller, 1912). Perhaps the relatively rare occurrence of these well-ordered clastic dike sets in the study area reflects a deeper setting in which displacements from surface waves were less pronounced, and tensional stresses were more subdued.

Features with less regular forms constitute a second variety of discordant liquefaction-injection phenomenon (Fig. 40). These are more widely distributed than their nearly planar counterparts, but are generally recognizable in the field only where textural or color contrasts distinguish the host rock from the clastic filling. Some are web-like networks of anastomosing veins or dikelets with widths of individual strands ranging from <0.1 mm to >3 cm. Excellent examples are found in the chaotic Twtb unit (Fig. 40a, specimen UW1837-17) in which irregular inclusions of green mudstone, siltstone, and sandstone host variably oriented

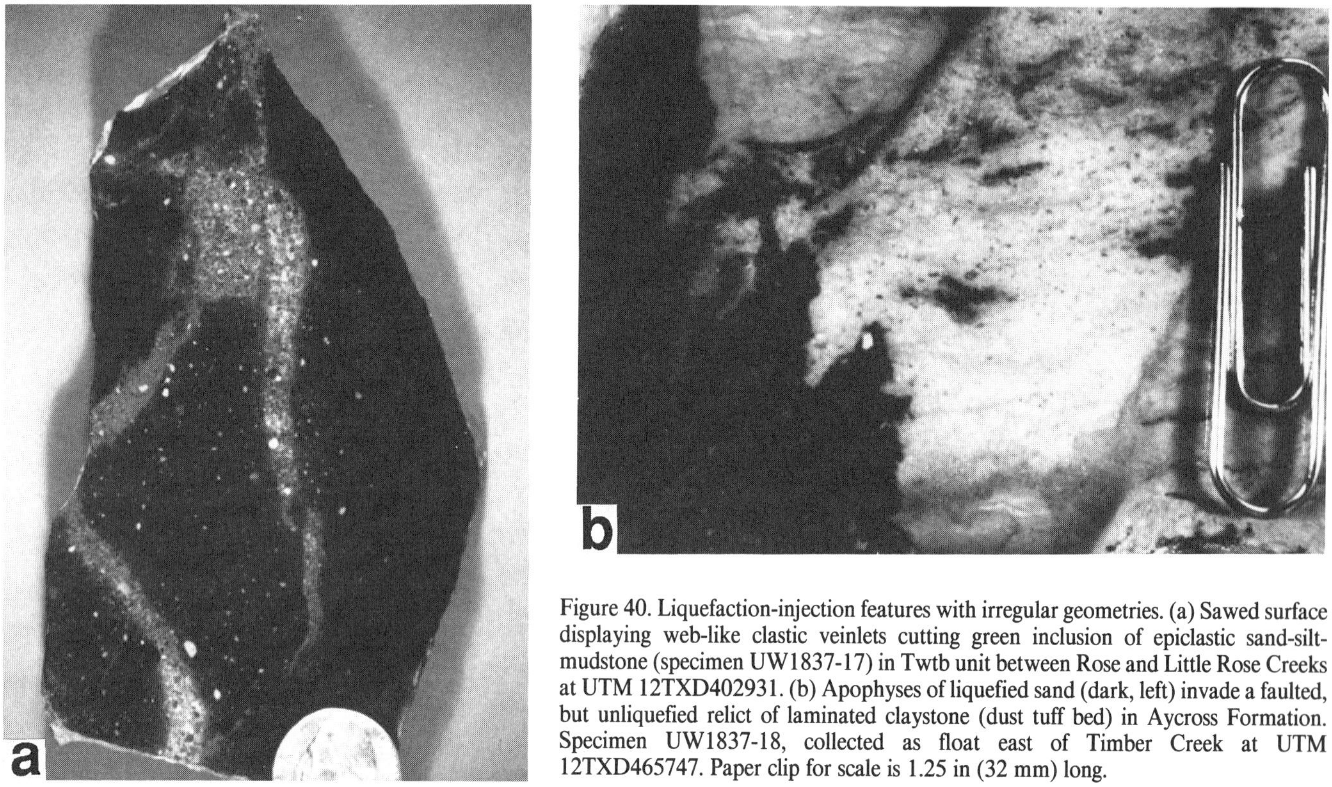

Figure 40. Liquefaction-injection features with irregular geometries. (a) Sawed surface displaying web-like clastic veinlets cutting green inclusion of epiclastic sand-silt-mudstone (specimen UW1837-17) in Twtb unit between Rose and Little Rose Creeks at UTM 12TXD402931. (b) Apophyses of liquefied sand (dark, left) invade a faulted, but unliquefied relict of laminated claystone (dust tuff bed) in Aycross Formation. Specimen UW1837-18, collected as float east of Timber Creek at UTM 12TXD465747. Paper clip for scale is 1.25 in (32 mm) long.

clastic veins filled with pale brown, fine- to coarse-grained sand. Others exhibit apophyses injected into faulted, but unliquefied sediments (Fig. 40b; specimen UW1837-18). These elaborate, irregular fissure systems may have formed by hydraulic fracturing and simultaneous filling of unliquefied units by the overpressurized pore fluids from neighboring beds.

Although injected clastic dikes are excellent evidence of liquefaction in overpressurized zones at depth, fluidization is actually the most probable mechanism of grain support (and, hence, liquidization) within the fissures themselves. The sediment having lost its strength by the development of fluid overpressures (liquefaction), grains become entrained in the escaping fluids, which carry them upward by viscous drag forces (fluidization). Hence, the mechanically distinct processes of liquefaction and fluidization may occur with some spatial overlap in a genetic continuum, with upward escape of fluids through fissures causing the transition between the two.

Dewatering pipes. Small ($\leqslant$1 cm diameter), crude cylinders of disturbed sediment occur locally in well-stratified beds of the lower Wapiti Formation (Twls unit). In this environment, they could be interpreted as dewatering pipes, degassing pathways from pyroclastic deposits, or possibly as traces of burrowing organisms. The best known examples occur in cross-laminated tuffaceous siltstones above the basal ignimbrite at the mouth of Houlihan Creek (Fig. 41; specimen UW1837-19). In most cases, these crudely cylindrical pipes are oblique or perpendicular to lamination, but their orientation varies considerably even at hand-specimen scale.

A larger cyclindrical feature in the Timber Creek drainage ($\sim$UTM 12TXD452759) also represents a possible relict pathway for the escape of overpressurized fluids. This vertical feature has a diameter of $\sim$20 cm, and is recognized by its slightly finer texture compared to the surrounding epiclastic breccia.

Ball-and-pillow structure and load structures. In the south-facing cliffs between Rose and Little Rose Creeks (UTM 12TXD401932), an approximately 200-ft-thick (60 m) by 800-ft-wide (240 m), synformal inclusion of andesitic conglomerate and conglomerate sandstone is completely engulfed by massive, homogeneous, Twtb unit breccias (Fig. 42). Beds at the outer edges of the inclusion are sharply upturned to dips that locally exceed 90°. As discussed above, the Twtb breccias possess many traits that suggest post-depositional reactivation as liquified masses. On the basis of its broad, sharply-upturned form and its isolation within the Twtb breccias, this inclusion is interpreted, despite its remarkable size, as an excellent example of ball-and-pillow structure. This constitutes the only known occurrence of ball-and-pillow structure in Aycross and lower Wapiti Formation rocks in the study area.

The high permeability of the conglomeratic strata within this pillow apparently precluded the development of fluid overpressures in this interval, despite the liquefaction of finer-grained epiclastic breccias surrounding it. This effect has been invoked (Wong and others, 1975) to explain the marked strain contrast between saturated gravels and liquefied sands involved in the 1964 Alaska earthquake. The sharply upturned bedding at the pillow's margins reflects vertical drag between the pillow and the

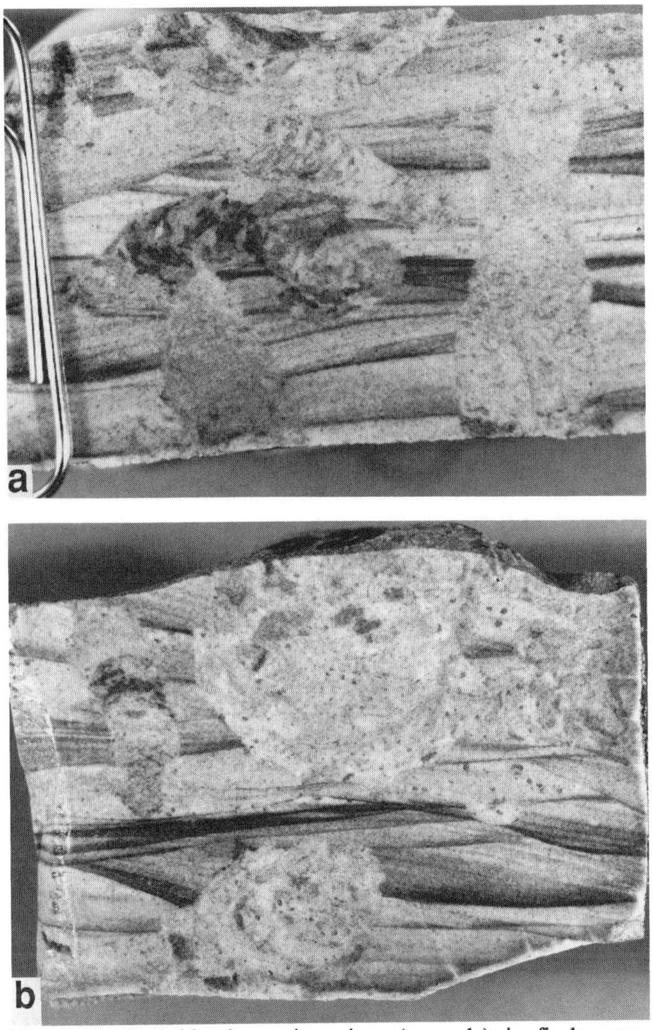

Figure 41. Probable dewatering pipes (cm-scale) in finely cross-laminated tuffaceous siltstone of Twls unit. Specimen UW1837-19, collected as float in Houlihan Creek at UTM 12TXE195065. Paper clip for scale is 1.25 in (32 mm) long; scale is the same in both views of parallel cuts through different parts of same specimen.

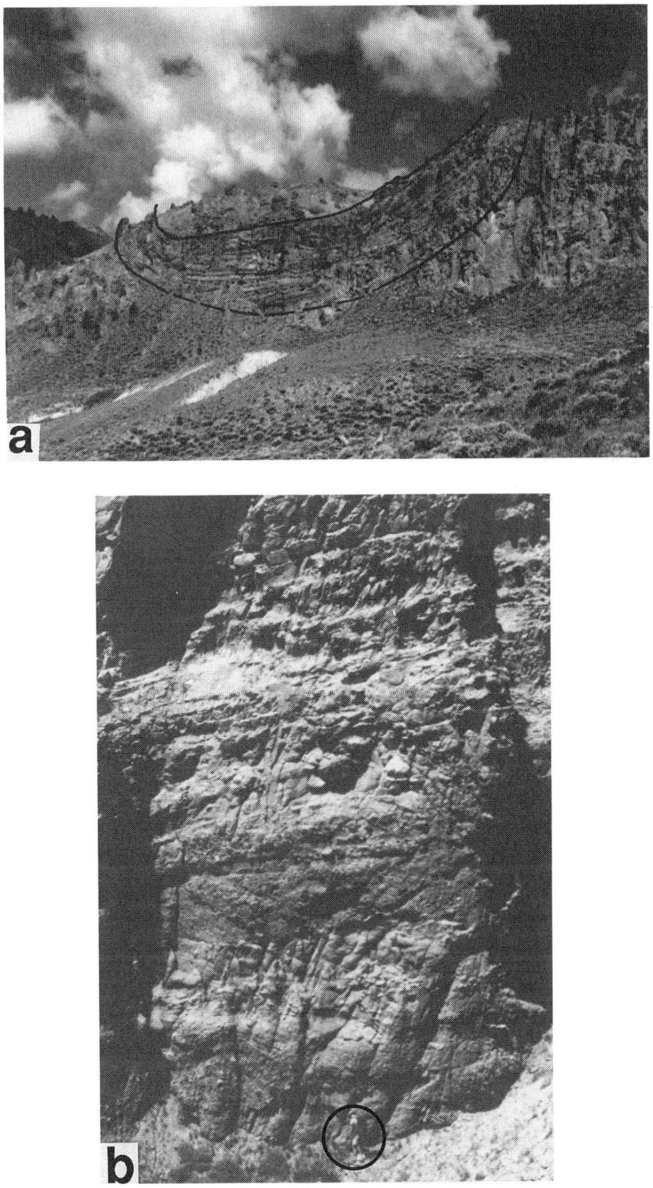

Figure 42. Relict pod of unliquefied Twca unit conglomerates within homogenized Twtb unit liquefaction breccias. (a) The inclusion measures roughly 200 ft thick by 800 ft wide on cliff face, but its overall synformal shape is that of ball-and-pillow structures commonly produced by liquefaction on a much smaller scale. Note sharply upturned beds at margins of pillow. (b) Detail of pillow's internal stratigraphic and sedimentologic character. Note trough cross-bedding in conglomerate strata near center, and geologist at base of cliff for scale. The coarse texture (high permeability) of these strata probably allowed pore fluid pressures to dissipate before reaching critical levels required for liquefaction. This pillow is represented schematically in the measured stratigraphic section of Plate VI.A.2. Both views are to northwest between Rose and Little Rose Creeks at UTM 12TXD4093.

surrounding liquefied material, probably caused by both upward motion of the liquefaction breccias and foundering of the relatively dense relict pod into the liquefied slurry.

Load casts are bulbous projections of denser, usually coarser-grained sediments into less dense, finer-grained, underlying beds. Like ball-and-pillow structures, load casts result from the differential vertical movement between strata of different viscosities, but are the product of less intense movement and, as such, are more common in these units than ball-and-pillow structures. Load casts alone are not necessarily indicative of liquefaction, but such a genetic relation is likely when they are found on the lower surface of stratified beds overlying a massive homogenized interval, as in Figure 39. Other examples in the study area range in wavelength from 1 to ~35 cm, and in amplitude from 2

Figure 43. Deformed contact in Aycross Formation epiclastic rocks shows attributes of both load casts and flame structures. However, the anvil-like shape was produced by "bulk exchange" of contrasting sediment types across the interface, rather than by penetrative interparticle slip. Locality is just north of Dick Creek at ~UTM 12TXD515724. Scale is 20 cm long.

mm to ~25 cm. Many of these are not juxtaposed upon liquefied sediment, and probably developed by normal consolidation of sediments of different densities.

The deformed contact of Figure 43 shows attributes of both load casts and flame structures, but is not a classic example of either one. Contrasts in texture, density, and viscosity between the light and dark sediments have produced anvil-like interpenetrations across the boundary. In deformations of this "bulk exchange" style, penetrative intergranular sliding is apparently limited to a rather thin boundary zone, which includes the delicate projections into the contrasting sediment, while the rest of the exchanged mass remains internally coherent.

Irregularly shaped inclusions. In addition to the orderly ball-and-pillow structures described above, Type 4 domains also contain prominent, irregularly shaped inclusions ranging in size from a few centimeters to several tens of meters across. Their delicate projections and awkward shapes are incompatible with derivation by surficial erosional processes, and would not likely have survived sediment transport across any significant distance. Rather, these inclusions are interpreted to have originated essentially in situ by dismemberment of more continuous preexisting strata. According to this model, although the textural and cohesive properties of some beds rendered them resistant to liquefaction, many beds were nonetheless intensely disrupted by the movements of sub- and superjacent liquefied material. The resulting inclusions of relict strata include many sizes of single-bed and multiple-bed pods; some are as much as 500 ft or more in length. The scale-independence of these and many other structures in the area is characteristic of deformation by liquefaction and related processes.

Figure 44 depicts an irregular inclusion of dark green volcanic sandstone exposed near the top of the liquefaction domain between Rose and Little Rose Creeks. The feathered edges of this and other nearby fragments of the same material indicate clearly their origin by disruption of a once-continuous bed or beds during liquefaction. Additional evidence supporting this interpretation is the absence of stratification in the enclosing sandstone, the diapiric invasion of the overlying stratified (and apparently unliquefied) sand by the massive sand, and the chaotic disarrangement of numerous inclusions and incoherent folds at the same locality.

Much larger inclusions, probably of similar origin, occur in the liquefaction interval between Piney and Pickett Creeks (Fig. 45; Plate VII.B). Measuring as much as perhaps 750 ft (230 m) in length (Hay, 1954), these multi-bed inclusions preserve their stratification, with only the edges of blocks being penetratively deformed. Hay (1954, p. 612) asserted that the bulk of one of these large blocks is structurally inverted. However, inspection of that block during this research revealed only a few inconclusive geopetal indicators (crossbeds with unclear sense of truncation, possible flaser bedding), and this interpretation remains suspect. Most of these inclusions are less than 200 ft (60 m) long (Fig. 45), and are surrounded by sedimentary intervals that show a progressive decrease in homogenization with decreasing depth of deformation.

All of the preceding examples document local resposnes to the widespread liquefaction of laterally equivalent units within the Aycross and lower Wapiti Formations. All of the deformation at this stratigraphic level probably occurred over a brief interval of time, perhaps even during a single event. However, there is evidence to suggest that liquefaction may have occurred

Figure 44. Deformational inclusion of green siltstone and sandstone formed by post-depositional strata disruption accompanying large-scale liquefaction. Note the minor diapirism (arrows represent probably movement) of liquefied sediment into the overlying unliquefied sandstone at top of photo. These intricately shaped green inclusions are common near the top of the Twtb unit liquefaction breccias in exposures throughout the east-central Absaroka Range, and they probably represent a formerly widespread stratum. View approximately east between Rose and Little Rose Creeks at UTM 12TXD401933, on the traverse of the measured stratigraphic section of Plate VI.A.2. Field notebook for scale.

Figure 45. Immense, irregularly shaped, internally stratified epiclastic inclusion in Wapiti Formation near east end of cliffs at Piney Creek–Pickett Creek Type 4 domain (~UTM 12TXD337896). For scale, a geologist is circled in dark recess near lower left corner of inclusion; diameter of white circle represents approximately 10 ft. Stratification is absent entirely in the homogeneous sandy liquefaction breccias that engulf the inclusion. The superjacent section is characterized by a complex intercalation of bedded remnants and homogenized intervals, and it becomes progressively less deformed upward toward the rigid lava flows and flow breccias of the overlying Twf/Twfb unit.

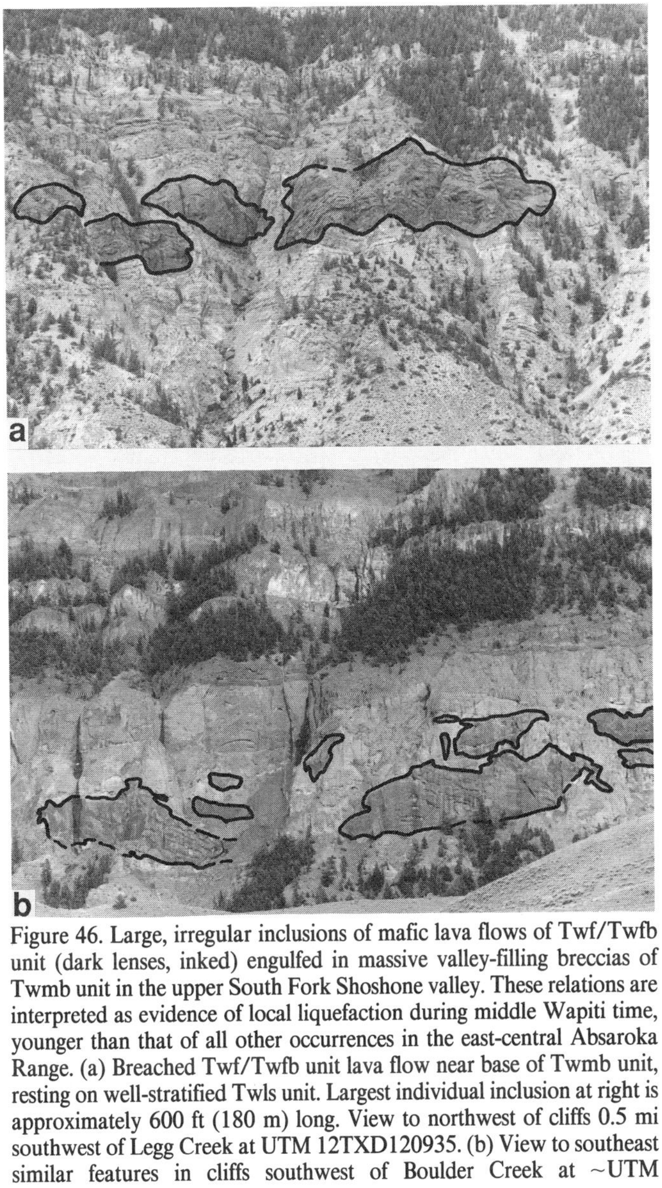

Figure 46. Large, irregular inclusions of mafic lava flows of Twf/Twfb unit (dark lenses, inked) engulfed in massive valley-filling breccias of Twmb unit in the upper South Fork Shoshone valley. These relations are interpreted as evidence of local liquefaction during middle Wapiti time, younger than that of all other occurrences in the east-central Absaroka Range. (a) Breached Twf/Twfb unit lava flow near base of Twmb unit, resting on well-stratified Twls unit. Largest individual inclusion at right is approximately 600 ft (180 m) long. View to northwest of cliffs 0.5 mi southwest of Legg Creek at UTM 12TXD120935. (b) View to southeast similar features in cliffs southwest of Boulder Creek at ~UTM 12TXD1591.

during middle Wapiti time as well, although on a much more local scale, in the upper South Fork Shoshone valley.

In the cliffs between Boulder and Hunter Creeks, and between Schoolhouse and Deer Creeks, on the southeast and northwest sides of the South Fork Shoshone valley, respectively, dozens of irregularly shaped pods of mafic lavas (Twf/Twfb unit)—the largest reaching 600 ft (180 m) long by 200 ft (60 m) thick—lie at a common horizon within the massive valley-filling Twmb unit breccias (Fig. 46). The features at this locality are unusual in at least two respects: the inclusions occur in a hosting unit (Twmb) younger than any other; and they are peperitic mixtures of sediments and magma that remained plastic enough to form intricate shapes upon incorporation into the enclosing sediments. At least two possibilities can explain these distinctive traits. First, the masses of autobrecciated lava may have intruded into the wet, unconsolidated Twmb unit breccias. Smedes (1966, p. 42–46) described strikingly similar occurrences in the Elkhorn Mountain Volcanics of Montana, and interpreted the irregular bodies as having intruded into the surrounding sediments. Alternatively, the masses may have formed by disruption and incorporation of a lava flow. The impulsive load delivered by lava-flow emplacement may have triggered local liquefaction of underlying saturated sediment. This scenario would conclude with the lava flow settling into the slurry, presumably losing its coherence and assuming the ragged forms shown in Figure 46. Such a local liquefaction event would be considered mechanically and temporally separate from the liquefaction-related deformation found elsewhere in the study area, which was probably triggered by cyclic loading during earthquakes.

Liquefaction diapirs. Hay (1954) noted several examples of cross-cutting relations between "intrusive tuff-breccia" and adjacent stratified intervals in the area of this study. Among the structures in the highly complex area between Rose and Little Rose Creek, Hay (1954, p. 611) described "breccia pipes and very irregular intrusive masses from 1 to 4 feet in diameter and as much as 10 feet long. The intrusions terminate upward within the bedded sequence. . . . The breccia of these small intrusive masses is similar to that of the breccia beds within the stratified unit and was almost certainly derived from them, probably by squeezing or other deformation while the beds were still unconsolidated." At Rose Butte, the chaotic, massive Twtb unit breccias cut diagonally across a section of stratified Twca unit andesitic conglomer-

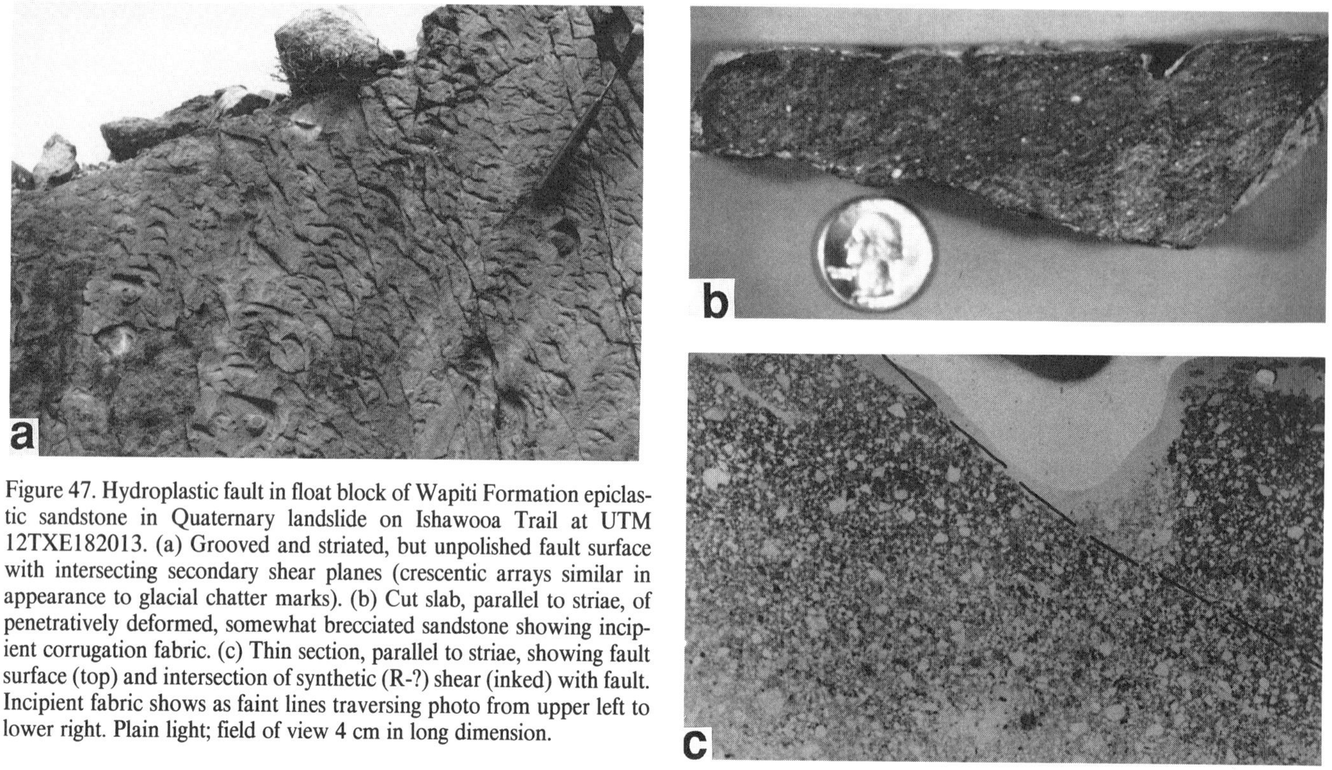

Figure 47. Hydroplastic fault in float block of Wapiti Formation epiclastic sandstone in Quaternary landslide on Ishawooa Trail at UTM 12TXE182013. (a) Grooved and striated, but unpolished fault surface with intersecting secondary shear planes (crescentic arrays similar in appearance to glacial chatter marks). (b) Cut slab, parallel to striae, of penetratively deformed, somewhat brecciated sandstone showing incipient corrugation fabric. (c) Thin section, parallel to striae, showing fault surface (top) and intersection of synthetic (R-?) shear (inked) with fault. Incipient fabric shows as faint lines traversing photo from upper left to lower right. Plain light; field of view 4 cm in long dimension.

ates (Fig. 6c; Plate VIII.A). The Twtb unit breccias lie beneath the Twca unit conglomerates in exposures on the northwest-facing slope of the butte, but overlie the same conglomerate beds on the knob south of the butte's summit. These relations are taken as evidence that the Twtb unit breccias moved diapirically at least 175 ft (50 m) through the Twca beds in a liquefied state. Similar liquefaction diapirism may have occurred at the top of the Type 4 domain immediately west of Hardpan Creek, where a bulbous mass of Twtb breccia extends upward into disheveled, but unliquefied, Twls unit strata. Structural relief on this bulbous mass reaches approximately 225 ft (70 m) (Fig. 24). Much smaller diapirs are common at contacts between massive, evidently liquefied, and stratified, unliquefied materials, as at the top of Figure 44.

Hydroplastic faults and deformation bands. Hydroplastic slickensides are striated surfaces formed by faulting of incompletely lithified sediments. They have been recognized only recently in the literature (Laville and Petit, 1984; Petit and Laville, 1987), and described on the basis of their morphologic and microstructural characteristics. Features that distinguish hydroplastic slickensides from fault surfaces formed in completely lithified rocks (Figs. 47 and 48) include: (1) a lack of frictional polishing; (2) arrays of crescentic patterns formed by the intersection of secondary (riedel) shear planes with the mullioned fault; (3) short, irregularly tapered grooves; (4) conjugate fractures symmetric about the transport direction; and (5) incipient fabric development related to cataclasis and offset along minor synthetic shears (specimen UW1937-20) (Petit and Laville, 1987). All of these traits are present in the known hydroplastic faults in the study area. Their limited occurrence in the intensely deformed Twls unit of the lower Wapiti Formation (Fig. 47) is significant because it bespeaks local variability in the extent of diagenesis undergone prior to middle Eocene deformation. Specifically, the Twls sandstones that exhibit hydroplastic slickensides must have been either somewhat more compacted or more cemented, or both, during deformation than those that experienced liquefaction. Even though this variability of sedimentological and mechanical characteristics is to be expected at burial depths of 100 to 200 m in a volcaniclastic pile, hydroplastic slickensides provide important documentation of the heterogeneity of deformational conditions.

As is generally true of other types of near-surface faults, the principal displacement on hydroplastic faults is confined to a surface rather than distributed across a zone. While hydroplastic faulting is, in this respect, physically different from liquefaction, it can occur within relatively more coherent strata as a direct response to the movement of liquefied material nearby. This mechanical link is exemplified by the intimate interaction between hydroplastic faults and fluid escape structures described by Guiraud and Séguret (1987) in fluviodeltaic cyclothems of the Soria Basin in northern Spain.

The distinctive morphology of hydroplastic slikensides makes them potentially useful as kinematic indicators (Laville and Petit, 1984; Petit and Laville, 1987). Unfortunately, the only known occurrences in the study area are from displaced and rotated blocks within Quaternary landslide masses, and recon-

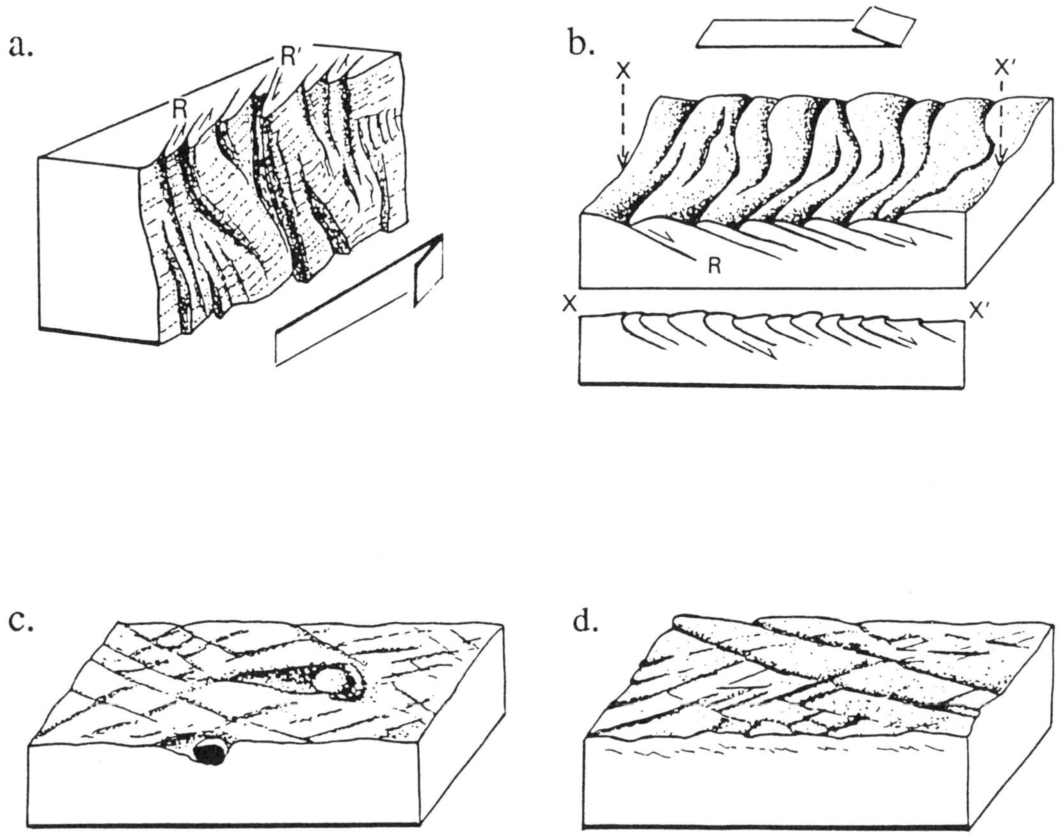

Figure 48. Morphologic characteristics of hydroplastic faults, simplified from Petit and Laville (1987). Compare with Figure 47. (a) and (b) Sketches of fault surfaces and movement-parallel sections show three-dimensional patterns of synthetic (R-) shear development. (c) Short, tapered grooves are made by gouging action of coarse fragments under a top-to-right sense of movement, and are filled in behind tool by viscous flow of unlithified sediments. Also note conjugate fractures distributed symmetrically about transport direction. (d) Conjugate fractures (surface), and incipient corrugation fabric (section). Shear sense is the same in (b), (c), and (d).

struction of middle Eocene liquefaction or sliding kinematics based on this morphology is not possible. Some of the gently dipping truncations of Type 3 domains (Figs. 21 through 25; Plate VIII.B) are probably hydroplastic faults, but diagnostic morphologic features have not been observed. Further investigation of Type 3 truncations—whether of thrust, low-angle normal, or presently indeterminate offset—could potentially reveal much about the kinematics and mechanics of deformation as well as the variability of diagenetic conditions. However, exposures of these surfaces are typically small, sparsely distributed, and difficult to reach.

Figure 49 shows a set of anastomosing dark bands in a highly silicified, tuffaceous sandstone (specimen UW1837-21). This sandstone occurs in a concordant zone (bed?) within moderately dipping beds on the eastern limb of the synformally deformed relict pillow of andesitic conglomerate shown in Figure 42. The outer surfaces of this 1 to 15-cm-thick zone are polished and striated by slip. In thin section, the irregular (0.5 to 10 mm wide) dark seams are found to represent zones of reduced grain size, containing fine grains of quartz, feldspars, muscovite, and hornblende. Platy and elongate grains within each seam are aligned subparallel to its trend. At least two possibilities exist to explain these observations. Inequant grains in the thin, dark seams may have been oriented by layer-parallel shearing during the synformal bending of these beds in the relict conglomeratic pillow while it remained unlithified. Limited cataclasis associated with this slip might account for the diminished grain size within the seams. Minor offset along anastomosing strands in the deformed zone would also explain the juxtaposition of sandy lenses with slightly different textures. This model is similar in many respects to the slip surfaces and zones of deformation bands described and illustrated by Aydin and Johnson (1983, Fig. 1c). The unusual hardness of the sheared zone may reflect the combined effects of strain hardening (Aydin and Johnson, 1983) and cementration by silica-rich fluids preferentially channeled along this formerly permeable, nearly clay-free zone.

Alternatively, grain reorientation may have been caused by overpressurized pore fluids forced through this permeable zone, particularly along these seams, during the liquefaction event that formed (and deformed) the pillow. However, in this case, elutria-

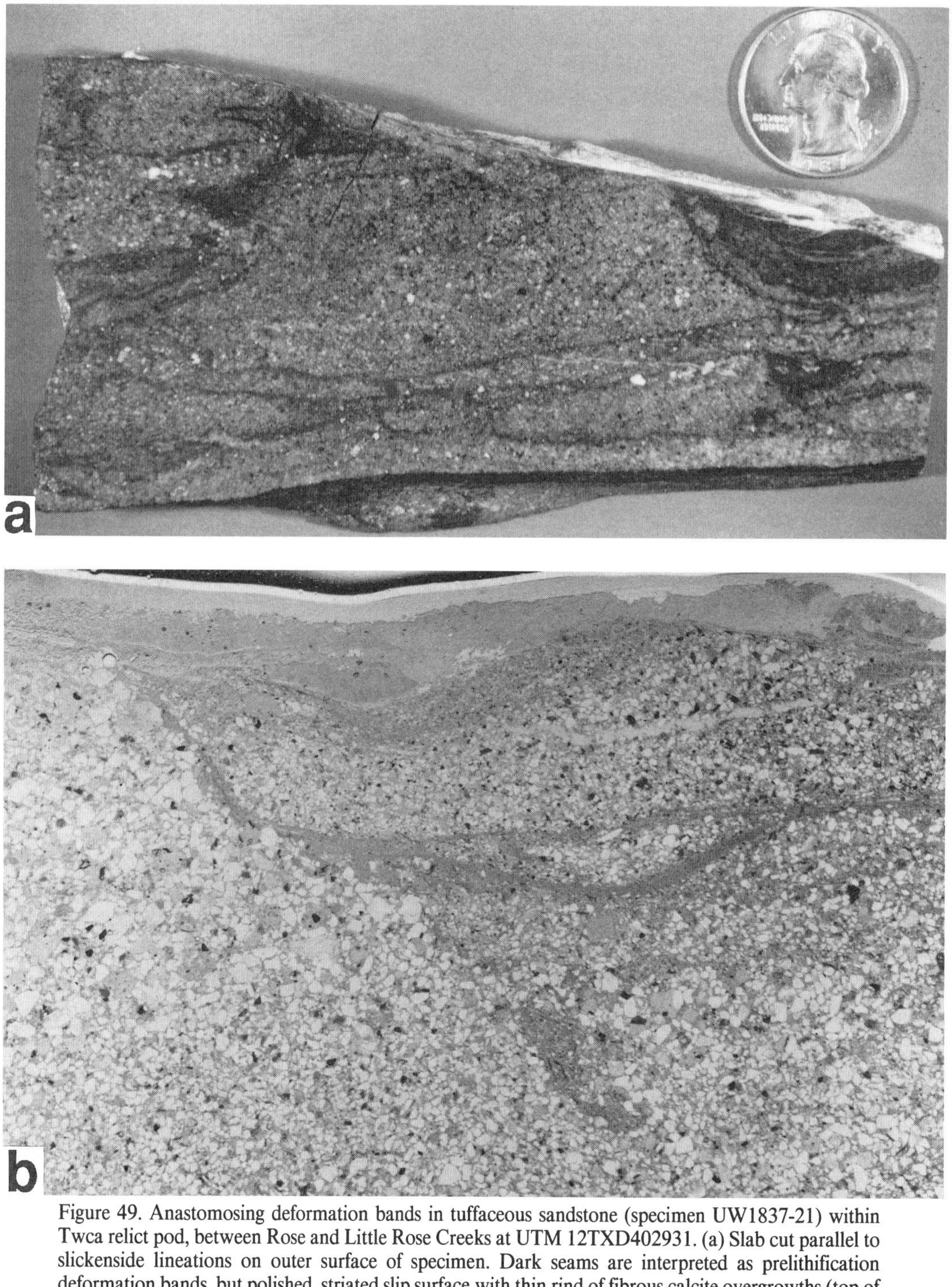

Figure 49. Anastomosing deformation bands in tuffaceous sandstone (specimen UW1837-21) within Twca relict pod, between Rose and Little Rose Creeks at UTM 12TXD402931. (a) Slab cut parallel to slickenside lineations on outer surface of specimen. Dark seams are interpreted as prelithification deformation bands, but polished, striated slip surface with thin rind of fibrous calcite overgrowths (top of specimen) probably developed after strain hardening or lithification. (b) Thin section cut parallel to slickenside lineations. Note slight textural differences on opposite sides of deformation bands. Plain light; field of view 4 cm in long dimension.

tion might be expected to have carried the finer sediments out of the fluid pathways, leaving them coarser rather than finer in comparison to the adjacent sediment. The former interpretation is favored over the latter, primarily because of this logical flaw. In the language of Aydin and Johnson (1983, Fig. 2), the anastomosing seams are taken to compose a zone of deformation bands bounded by relatively late-forming, striated slip surfaces.

Conditions of deformation

The present stratigraphic and structural configurations of some Type 4 domains provide enough information to reconstruct approximately the conditions of deformation by liquefaction and related dismemberment processes. The most significant factors controlling this deformation, including depth of burial, lithostatic stress, pore fluid pressure, degree of hydraulic confinement, and the homogeneity of deformation, can be inferred at optimally exposed localities in the study area (Decker, 1988). If it is assumed that this regionally extensive liquefaction was triggered mainly by seismic activity, probable earthquake magnitudes can also be estimated using an empirical formula that relates seismic intensity to the size of the area experiencing liquefaction (Kuribayashi and Tatsuoka, 1975).

The two Type 4 domains most successfully analyzed for local conditions of deformation are the Piney Creek–Pickett Creek and Hardpan Creek domains. Here, an angular unconformity is developed at the top of the involved interval, and the minimum depth of burial during deformation is fairly well constrained. The lithostatic stress at a point depends only on the depth and density of the overlying material. Hence, if approximate density values and stratigraphic proportions are known for the two principal types of overburden, lava flows and partially consolidated epiclastic sediments, lithostatic pressure can be estimated at these two localities. Furthermore, because pore fluid pressures approaching the lithostatic stress are required for liquefaction to occur, estimates of minimum critical pore pressures are equally feasible at these localities. However, only at the Cabin Creek–Deer Creek domain is the full thickness of the liquefied and chaotically dismembered interval defined by clear exposure of its lower contact (Plate VII.A). Hence, because liquefaction elsewhere may have extended to structural levels deeper than those currently exposed, maximum pore pressures (presumably developed at the base of the liquefied interval) may be underestimated at those localities.

At the Rose Creek–Little Rose Creek, Timber Creek, and Cabin Creek–Deer Creek domains, the depth of deformation is poorly constrained. At these localities, the conditions of deformation can only be qualitatively described in terms of the apparent presence or absence of a hydraulic confining layer, and the deformation's geometry, intensity, and pervasiveness.

Deformation at Piney Creek–Pickett Creek locality. As illustrated in Plate VII.B, the zone of liquefaction and intense stratal dismemberment between Piney Creek and Pickett Creek is overlain by an ~300-ft (90 m) sequence of tilted, faulted, and

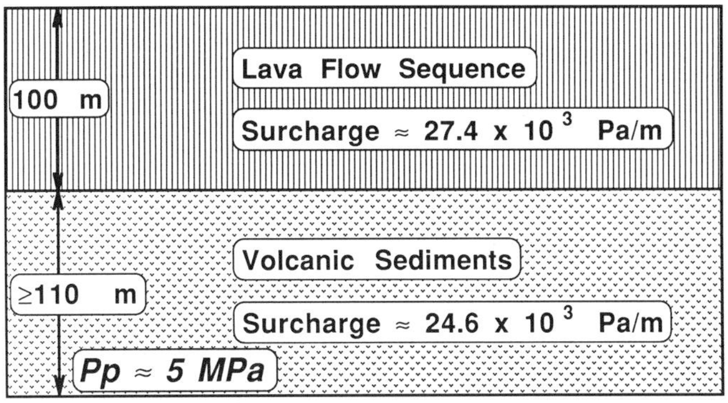

Model for Estimate of Pore Pressure at Pickett Creek -- Piney Creek Locality

Figure 50. Model for estimate of pore fluid pressure at Piney Creek–Pickett Creek locality. The lower Type 4 interval is part of Tw unit, overlain by essentially rigid lavas of Twf/Twfb unit. Refer to Plate VII.B for geologic detail. Pore fluid pressures probably reached at least ~5 MPa at base of Type 4 domain.

erosionally beveled lava flows. Near the western end of this spectacular exposure, the downdropped flow sequence displaces a subequal thickness of disrupted sediments. Near the center of the exposure, a block of autobreccia has been isolated and rotated 15°, and the preserved thickness of the lavas thins to zero in a poorly exposed section about 250 ft (75 m) in length. This area is interpreted in Plate VII.B. as a breach through which liquefied sediments were driven upward under buoyancy drive, simultaneously with the lowering of the denser flow sequence several hundred meters to the west. These relations clearly indicate that the flow sequence was involved in deformation. Moreover, it is likely that these flows exerted considerable control on this deformation by dramatically enhancing hydraulic confinement, which assists the pressurization of pore fluid, and by providing the density-inversion instability required for large-scale displacements of liquefied material.

Lithostatic and pore fluid pressures at the time of liquefaction can be estimated on the basis of the stratigraphic thicknesses and densities of the material in and above the liquefied zone using the simple relation

$$\sigma_v = P_f = \rho g h$$

where σ_v = vertical confining pressure (lithostatic stress), P_f = pore fluid pressure, ρ = density, g = gravitational aceleration, and h = depth of burial. At this locality, saturated densities are estimated (with consideration given to values provided by Carmichael, 1984), to be ~2.8 g/cm^3 for the mafic autobreccias of the ⩾300-ft (~100 m) flow sequence, and ~2.2 g/cm^3 for the partially consolidated clastic sediments of the ⩾360-ft (110 m), largely disrupted interval. These densities are readily converted into vertical pressure per unit thickness of overburden (surcharge) values: the surcharge due to the lava flows becomes 27.4 kPa/m and that from the clastic sediments becomes 24.6 kPa/m. On the basis of these thickness and overburden values (Fig. 50), the pore fluid pressure required for liquefaction must have reached at least 5 MPa (725 psi) at the base of this liquefied and dismembered interval.

The disrupted interval at this locality is by no means thoroughly homogenized by liquefaction. Relict inclusions of stratified material—as semi-rigid, multiple-bed pods and intensely contorted individual beds—make up the bulk of this exposure, but most are encased in the unstratified, more easily eroded liquefaction breccia that locally dominate the exposure.

Deformation at Hardpan Creek locality. The genetic relation between deformation of Type 4 domains and other structural styles is nowhere more apparent than at the Hardpan Creek locality (Fig. 24). Quaternary landsliding provides excellent exposures in south-facing cliffs and precipitous slopes with approximately 600 ft (180 m) of relief. Although the contact between the Type 4 liquefaction breccias and the overlying Type 3c domain, which contains truncation surfaces with unclear sense of movement, is very irregular, on the average, each domain constitutes roughly half of the total exposed thickness of the deformed sedimentary interval.

The saturated density for this interval is assumed uniform at 2.2 g/cm^3, the same value used for the clastic section at the Piney Creek–Pickett Creek locality. Figure 51 illustrates the schematic model used to calculate an approximately pore fluid pressure of at least 4 MPa at the base of the liquefied section. This is about 20 percent less than that estimated in the previous locality, where the minimum depth of deepest liquefaction is slightly greater, and where the relatively dense lavas contributed additional confining pressure. The sedimentary section at Hardpan Creek is incompletely preserved, and liquefaction there may have occurred at depths and pressures similar to or greater than those at the Piney Creek–Pickett Creek locality.

While there is no well-defined individual confining layer capping the Type 4 liquefaction domain at Hardpan Creek, the overlying Type 3c domain's vertical hydraulic conductivity was probably significantly reduced by the presence of thick mud- and silt-rich interbeds. The aquitard characteristics of this well-stratified sequence may have been lost as the numerous truncation surfaces (hydroplastic faults?) developed in response to the deeper liquefaction.

Causes of fluid overpressures. Pore fluids in unconsolidated sediments may be pressurized by either dynamic or static effects, or both. Dynamic pressurization occurs when external loading destroys the grain contacts, shifting the weight of the solid phase to the pore fluid. The cyclic loading associated with earthquake vibrations represents the most likely triggering mechanism for widespread liquefaction (Owen, 1987), and probably played a crucial role in the middle Eocene deformation of the Wapiti and Aycross Formations. However, impulsive liquefaction may be induced by the sudden deposition of lava flows or debris flow units, and this may have exerted at least local influence on deformation. Indeed, at the Piney Creek–Pickett Creek liquefaction locality, lava flows appear to be the youngest strata involved in deformation. A compressional front (possibly even true shock

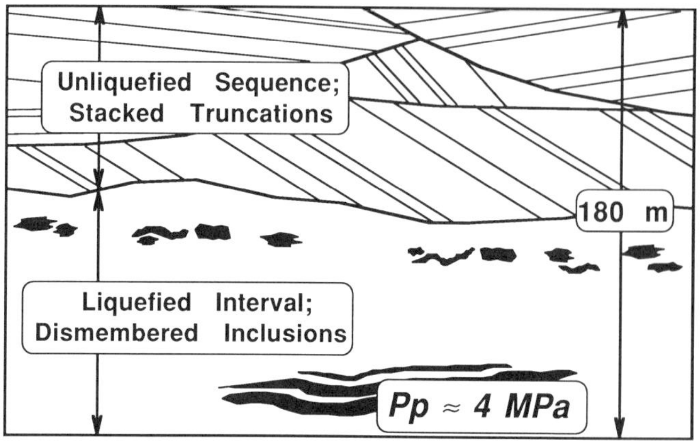

Figure 51. Model for estimate of pore fluid pressure at Hardpan Creek locality. The lower Type 4 interval is Twtb unit, overlain by Type 3 interval in Twls unit. Refer to Figure 24 for geologic detail. Pore fluid pressures probably reached at least ~4 MPa at base of Type 4 domain.

waves) radiating from an explosive Absarokan eruption represents yet another possible source of impulsive loads in this region.

Static effects may also have contributed to fluid overpressures. Aquathermal pressuring (Barker, 1972) is one means of developing zones of abnormally high pore fluid pressure by heating fixed volumes of isolated water. This process was originally recognized in the context of the heating that occurs as Gulf of Mexico shelf sediments subside through the isotherm field (Barker, 1972). However, the same mechanism may be effective in pressurizing trapped groundwater at shallow depths in regions of high heat flow, such as active volcanic terranes. The significance of this effect on elevated pore pressure causing liquefaction depends on the extent of hydraulic confinement and the amount of heating. Presumably, this type of pressurization would be most pronounced in the vicinity of intrusive masses.

There is no criterion for quantifying the possible contribution to pore fluid pressurization from each of these dynamic and static sources. Each may have been at least locally significant. The most striking characteristic of Wapiti/Aycross deformation is its semi-regional extent, which appears to indicate either repeated or prolonged seismic shaking or extremely violent explosive eruptions as the most likely triggers. There is no distinctive pyroclastic blanket preserved at stratigraphic level inferred to be at the surface during liquefaction. Lacking any other evidence linking deformation to explosive energy, seismic activity is favored as the most likely and most significant triggering mechanism. This early middle Eocene seismic activity was probably associated with Absaroka magmatism, but it may have been more directly related to Laramide compression, although this regime was by then waning markedly.

Estimate of earthquake magnitude. In their study of 44 Japanese earthquakes, Kuribayashi and Tatsuoka (1975) determined an empirical relation between the maximum epicentral distance of liquefaction sites R (km) and the earthquake magnitude M:

$$\log_{10}R = 0.77M - 3.6$$

This formula was tested by Tohno and Yasuda (1981), who found it to be consistent with the distribution of sites that experienced liquefaction during the magnitude 7.4 Miyagiken-Oki earthquake. While this simple formula does not consider the sediment's textural characteristics, and may not be strictly valid in tectonic settings other than that in which it was derived, it may be useful here to estimate the magnitude of Eocene earthquakes causing liquefaction in the east-central Absaroka Range.

The known liquefied intervals in the Aycross and lower Wapiti Formation are distributed in an area ~50 km from northwest to southeast. The most conservative estimate of earthquake magnitude would be obtained by placing the epicenter just southwest of Piney Creek, near its junction with the Greybull River, at a point equidistant from the limits of this known liquefaction. The epicentral radius would thus be 25 km, and the magnitude calculated from Kuribayashi and Tatsuoka's (1975) formula would be $M \approx 6.5$. Although this represents a strong earthquake, it is rather small for such intense liquefaction when compared to the observed effects of stronger historic earthquakes.

Heart Mountain faulting, which may be the same age as the liquefaction events considered here, may also be genetically related as a mechanically different response to the same seismic stimulus. If this is true, the most probable epicenter would be near the Heart Mountain fault in the northern Absaroka Range, rather than in the Greybull River drainage. The massive intrusive center and volcanic vent complex centered on Sunlight Peak stands out immediately as the most apparent source of seismic energy in the northern Absaroka Range. Based on a maximum epicentral distance of 85 km from the center of this complex to the most distant observed liquefaction site in the Dick Creek drainage, the calculated magnitude increases to $M \approx 7.2$, which agrees more closely with magnitudes of historic earthquakes known to have caused regionally extensive liquefaction. This value may be conservative still; it would increase upon verification of liquefaction sites farther south, with possible candidates being the "chaotic tuff breccia" localities described by Barnes (1985) in the Wood River area. Whatever the seismic magnitude may have been, the great thickness (100 to 200 m) of unlithified Absaroka sediments throughout the region enhanced the susceptibility to liquefaction (Nuttli, 1973).

Liquefaction and related deformation in historic earthquakes

New Madrid, Missouri; 1811–1812. In one of the most widely felt releases of seismic energy in historic times, the series of earthquakes in 1811–1812 distributed about New Madrid, Missouri, triggered a swarm of surface and subsurface deformations by liquefaction and related processes. Three principal shocks with estimated body wave magnitudes of 7.2, 7.1, and 7.4 (Nuttli, 1973) occurred between December 16, 1811, and February 7, 1812, accompanied by a terrifying series of aftershocks felt through December, 1813 (Fuller, 1912). Vibrations from the largest pulse were felt as far away as Middlebury, Vermont, some 1,000 mi (1,600 km) distant.

Deformational phenomena associated with seismicity included several styles of ground fissuring, faulting, landsliding, locally pronounced doming and subsidence, development of clastic dikes, extrusion of mixtures of water, gas, and sediment (sand blows, sand sloughs), and even the complete disappearance of islands in the Mississippi River (Fuller, 1912). The area of pervasive liquefaction, as recognized at the surface by extruded sediment, trends NNE–SSW in a rough ellipse about 140 km (88 mi) long and 50 km (31 mi) wide (Fuller, 1912, Plate I) and coincides closely with the thick accumulation of Mississippi River alluvium. It is precisely this ≥1,000-ft-thick (≥300 m) unconsolidated alluvial cover (Fuller, 1912), of which some interval was water saturated, that was responsible for the unusual severity of deformation in this region (Nuttli, 1973).

Liquefaction accompanying the New Madrid earthquakes is most clearly documented by eyewitness accounts of extrusive activity at "sand blows" and "sand sloughs" (Fuller, 1912; Penick, 1976), and by excavations that reveal upward- and laterally-injected clastic dikes (Fuller, 1912, Plate III and Fig. 8). In addition to these clear demonstrations of liquefied behavior, much landsliding (Fig. 52a), faulting, and possibly some ground warping also occurred as secondary responses to instabilities created by this liquefaction (Fuller, 1912). Except for sediment extrusions (which would be difficult to identify in the geologic record even if they survived the intra–Wapiti/Aycross erosional episode that followed deformation), these same effects are documented in the Aycross and lower Wapiti Formations in the Absaroka Range. Their proposed relation to liquefaction there finds firm support in just such a linkage during the Mississippi Valley earthquakes of 1811–1812.

Niigata, Japan; 1964. The magnitude 7.5 earthquake of June 16, 1964, in Niigata, Japan, ranks high among the most thoroughly studied seismic events ever. The intense destruction caused by this shock reflected the combined effects of vibration of structures, extensive liquefaction of the city's nearly horizontal alluvial substrate, fires that broke out at damaged oil storage tanks, and the accompanying tsunami (Kobayashi, 1968). Widespread ejection of sand and water added to the damage that resulted from remarkable foundation failures, which in a few cases, left whole structures essentially intact, although steeply inclined or rotated nearly to the horizontal (Kawasumi, 1968, Plates 7, 11). More commonly, liquefaction failure in level urban areas caused foundations to subside differentially, resulting in pervasive structural damage or collapse (Kobayashi, 1968; Seed and Idriss, 1982).

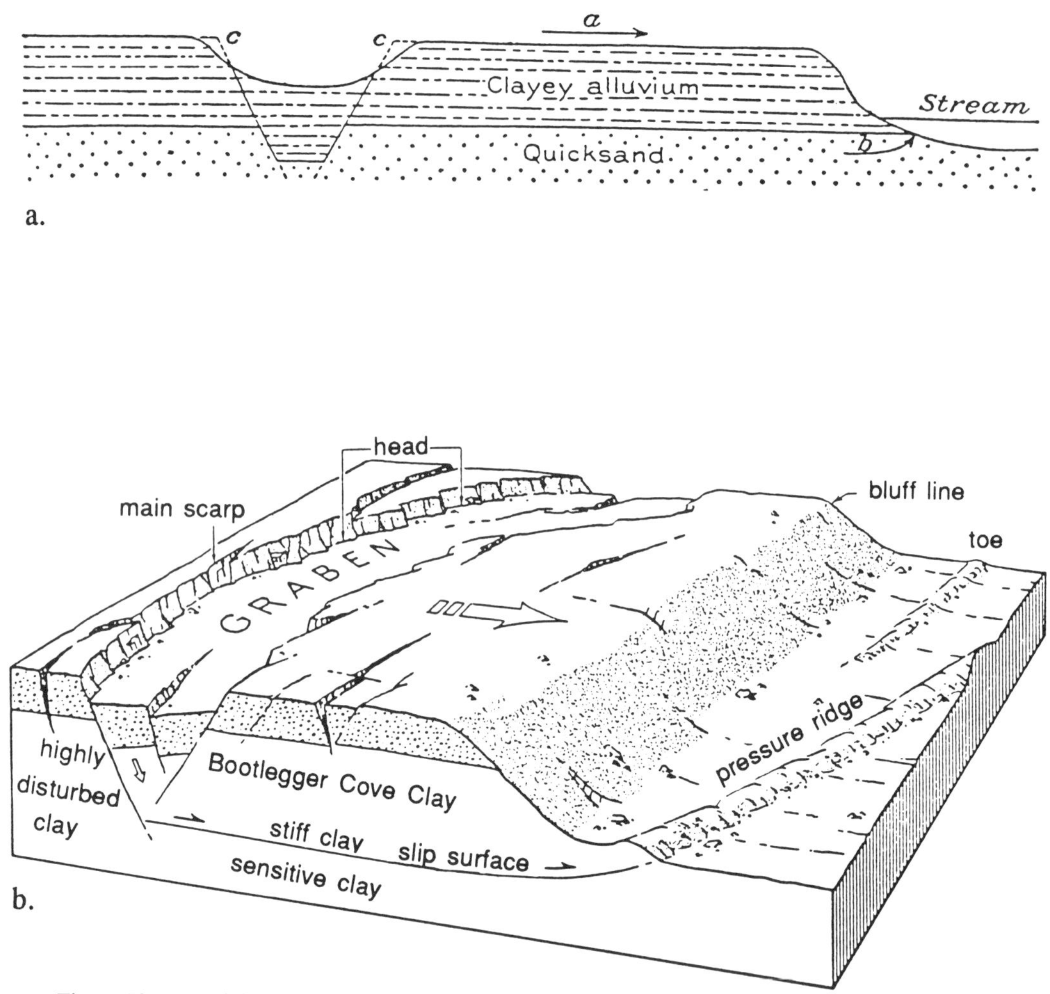

Figure 52. Landsliding as a response to documented liquefaction or other liquidization processes at depth. (a) In the New Madrid earthquakes of 1811–1812, essentially rigid, block-faulted slabs of surficial material moved laterally toward unconfined faces upon liquefaction of underlying alluvial sediments (quicksand). From Fuller (1912, Fig. 6). (b) Liquefaction coupled with disturbance of the sensitive Bootletter Cove clay substrate during the 1964 Alaska earthquake promoted similar lateral transport toward free faces. This mechanism was largely responsible for the catastrophic Turnagain Heights landslide. Modified from Hansen (1965, Fig. 24). Such liquefaction-induced landsliding probably occurred locally during early middle Eocene deformation in the Absaroka Range, particularly in association with listric normal faulting in Type 3a domains. In addition, some Type 3b domains (characterized by thrust faults) may represent the local compressional regime at the landslide-toe shown in (b).

Rural areas surrounding the city also saw extensive extrusion of liquefied sand. In addition, the less populated, largely unpaved land bore witness to other deformational phenomena such as ground fissuring, step faulting, and slope failure (Kawasumi, 1968, Plates 38, 39).

Ishihara and Koga (1981) combined soil sampling and testing with seismic acceleration records from the 1964 Niigata earthquake to compute profiles of factor of safety (susceptibility to failure) versus depth. They concluded that liquefaction occurred throughout a zone between 2 and 13 m (6 to 43 ft) depth at Kawagishi-cho, a site where large apartment buildings were tilted and overturned. In a site 1.2 km (0.75 mi) away, where no sand blows or other surficial indications of liquefaction were found, their data indicate that liquefaction may have occurred in a zone $\leqslant 1$ m thick at 4 to 5 m (12 to 15 ft) depth. This is significant, for it implies that even in very shallow environments, limited liquefaction may have no surface expression. In other words, liquefaction may cause deformation at depth well beyond the area in which its effects are observed at the surface.

South-central Alaska; 1964. The earthquake that shook more than 500,000 mi^2 of south-central Alaska on March 27, 1964, was unusual not only in magnitude (M = 8.3 to 8.6), but also in the duration of shaking (locally exceeding 4 min) (Seed and Wilson, 1967; Krauskopf, 1971). This uncommon combination proved disastrous across a heavily damaged area of some 50,000 mi^2 (Krauskopf, 1971), particularly in urbanized Anchorage, about 80 mi WNW of the epicenter.

Damage in the vicinity of Anchorage was due in large part to landsliding and other ground failures resulting from disturbance of a highly sensitive marine clay facies of the Bootlegger Cove Formation, and more local liquefaction of silt and sand interbeds within this unit, both of which were abetted by the strength and duration of the vibrations (Hansen, 1965; Seed and Wilson, 1967; Updike and others, 1988). As in the New Madrid earthquakes, liquidization of subsurface layers provoked catastrophic translatory landsliding of surficial units toward unconfined faces (Fig. 52b).

Although numerous smaller slides caused by the 1964 Alaska earthquake are equally demonstrative, the distressing spectacle of the Turnagain Heights landslide (Hansen, 1965; Seed and Wilson, 1967; Updike and others, 1988) has become the classic example of this process. Its complex internal structure bears local resemblance to the otherwise enigmatic stacked truncation surfaces, listric normal faults, and thrust faults of Type 3 domains in the AVS. While liquidization in Aycross/Wapiti sediments seems to have involved liquefaction nearly to the exclusion of thixotropism and sensitivity, the resultant mechanical effect on overlying strata may have been essentially the same as that which triggered large-scale failures in the Alaska earthquake.

Summary

Among the numerous styles of deformation in the Aycross and lower Wapiti Formations, liquefaction is considered here to be the most widespread and most significant in governing the structural development of the east-central Absaroka Range during early middle Eocene time. Type 4 domains contain a suite of structures produced by liquefaction, cyclic mobility, and complex stratal dismemberment, and they are closely associated with structural domains of the three other types. This relation suggests that the instabilities that gave rise to gravitational deformations of Types 1 through 3 at shallow levels were in turn spawned during adjustment to liquefaction at depths reaching 200 m or more.

Pore fluid pressures during deformation are estimated to have reached at least ~5 MPa at the locality of deepest known liquefaction. This condition was probably achieved in response to both static and dynamic factors associated with Absaroka volcanism. In particular, one or more earthquakes of magnitude $M \geqslant 6.5$—and more likely $M \geqslant 7.2$—probably represent the major stimulus for this liquefaction. Historic and recent earthquakes continue to provide valuable information with which to interpret the geologic record of subsurface deformations, such as those so strikingly exhibited in the AVS.

EARLY CENOZOIC DEFORMATIONAL HISTORY AND STRATIGRAPHIC DEVELOPMENT

The mountainous topography of the study area in the east-central Absaroka Range is carved nearly entirely in Tertiary rocks deposited following the main Laramide orogenic phase of Paleocene time. Only in the floors of the South Fork Shoshone and Greybull valleys are Mesozoic rocks exposed. The sequence of Cenozoic events that gives these mountains their present geologic configuration includes the episodes of volcanic and nonvolcanic deposition, intrusive activity, deformation, and erosion summarized in Table 6.

Early Eocene

Laramide folding culminated during Paleocene time, accompanied by erosional stripping throughout the study area. This gave way to fluvial and fluviolacustrine deposition (Van Houten, 1944; Bown, 1980; Kraus, 1985) of the lower Eocene Willwood Formation, which caps the main Laramide unconformity with strong angular discordance. Fluvial channeling of this unconformity surface is locally evident in the study area, as at Red Butte near Timber Creek in the Greybull Valley (UTM 12TXD471816).

Both the Willwood Formation and the underlying angular unconformity locally are gently arched over the core of the NW-SE–trending Laramide Fourbear–Willow Draw anticline. Because AVS units have been stripped from the anticlinal axis by erosion, it is unclear whether they were also involved in this post-Willwood folding; hence, both the youngest possible age and the process of latest folding there remain enigmatic. If only the Willwood and older formations were affected, warping must have occurred in late early Eocene time, and a brief pulse of latest Laramide compression may simply have reactivated the anticline. However, if younger AVS units were also warped here, as they are along a parallel trend nearby to the southwest (beneath Irish Rock and Phelps Mountain; Fig. 53), this warping may have been caused by intrusion of an unexposed dacite body (known from drilling data) into the core of the Laramide anticline (Fig. 3), and may be of late Eocene or younger age. Radiometric dating of the subsurface dacite intrusion(s) in the Greybull River valley would be valuable in resolving this issue.

Early middle Eocene

Silicic to intermediate pyroclastic deposits and their tuffaceous epiclastic counterparts at the base of the AVS dominated the fluvial environment of the area by early middle Eocene time. Trachyandesitic lavas and laharic debris flows also became locally significant in this interval. With some major exceptions, these strata were deposited in rather well bedded sections recognized here as the Twls, Twtb, Tws, Twca, and Twf/Twfb units, as well as parts of the Tw, Ta, and Twlb units of the lower Wapiti and Aycross Formations.

These units were locally intensely deformed during early middle Eocene time, following deposition and burial to depths reaching at least 200 m (~650 ft), but prior to extensive lithification. Widespread instabilities induced by pore fluid overpressures

TABLE 6. SUMMARY OF TERTIARY DEFORMATIONAL HISTORY AND STRATIGRAPHIC DEVELOPMENT IN THE EAST-CENTRAL ABSAROKA RANGE

Time Interval	Rock Units	Event(s) and/or Dominant Process
late Eocene or more recent	upper Twi or younger (no post-Twi section)	Gentle folding about NW-SE axes, normal faulting
middle to late Eocene	Ttp, Tbp, Tt, Twi (upper AVS units)	Migration of volcanic centers to SSE with deposition of upper AVS units, emplacement of andesitic radial dikes, emplacement of exposed dacitic intrusions and associated mineralization
middle Eocene	middle and upper units of Wapiti & Aycross Fms. (Twj, Twmb, Twu, Twuf, Tas, remainder of Tw, Ta)	Burial of landscape forms intraformational unconformity, massive breccia and sand units deposited by sediment-water flows, lava flows of Jim Mountain Member and upper Wapiti Fm.
middle Eocene	(after deposition and deformation of lowest AVS units; no section)	Erosion: •truncation of deformed intervals •locally great paleotopographic relief
early middle Eocene	(after deposition, but prior to complete lithification of lower Wapiti & Aycross units)	Prelithification deformation: •Type 4 domains: liquefaction & cyclic mobility, intense dismemberment •Type 1, 2, and 3 domains: responses to liquefaction-induced instabilities Local prelithification intrusion of early dacitic dikes and sills
early middle Eocene	lower units of Wapiti & Aycross Fms. (Twls, Twtb, Tws, Twca,Twf/Twfb, parts of Tw, Ta, Twlb)	Early explosive eruptions, fluvial and fluviolacustrine deposition, laharic debris flows, limited lava flows
(?) late early Eocene to earliest middle Eocene *	(post-Willwood Fm., pre-AVS; no section)	Latest-stage Laramide compressional or intrusive folding of Willwood & underlying Laramide angular unconformity*
early Eocene	Willwood Fm.	Burial of main Laramide angular unconformity, fluvial and fluviolacustrine deposition
Paleocene	(no Paleocene section)	Major Laramide folding, erosion

* Partly because of erosion of AVS units in floor of Greybull RiverValley, neither timing nor process of gentle folding of Willwood Formation and underlying Laramide angular unconformity is adequately constrained. If folding involved only Willwood and older units, it must have occurred in late early Eocene time, and a latest-stage Laramide compressional origin is favored. If AVS units were also involved, folding may have been caused by intrusion of dacite body into core of Four Bear--Willow Draw anticline (see Fig. 3), which may have occurred as recently as late Eocene time.

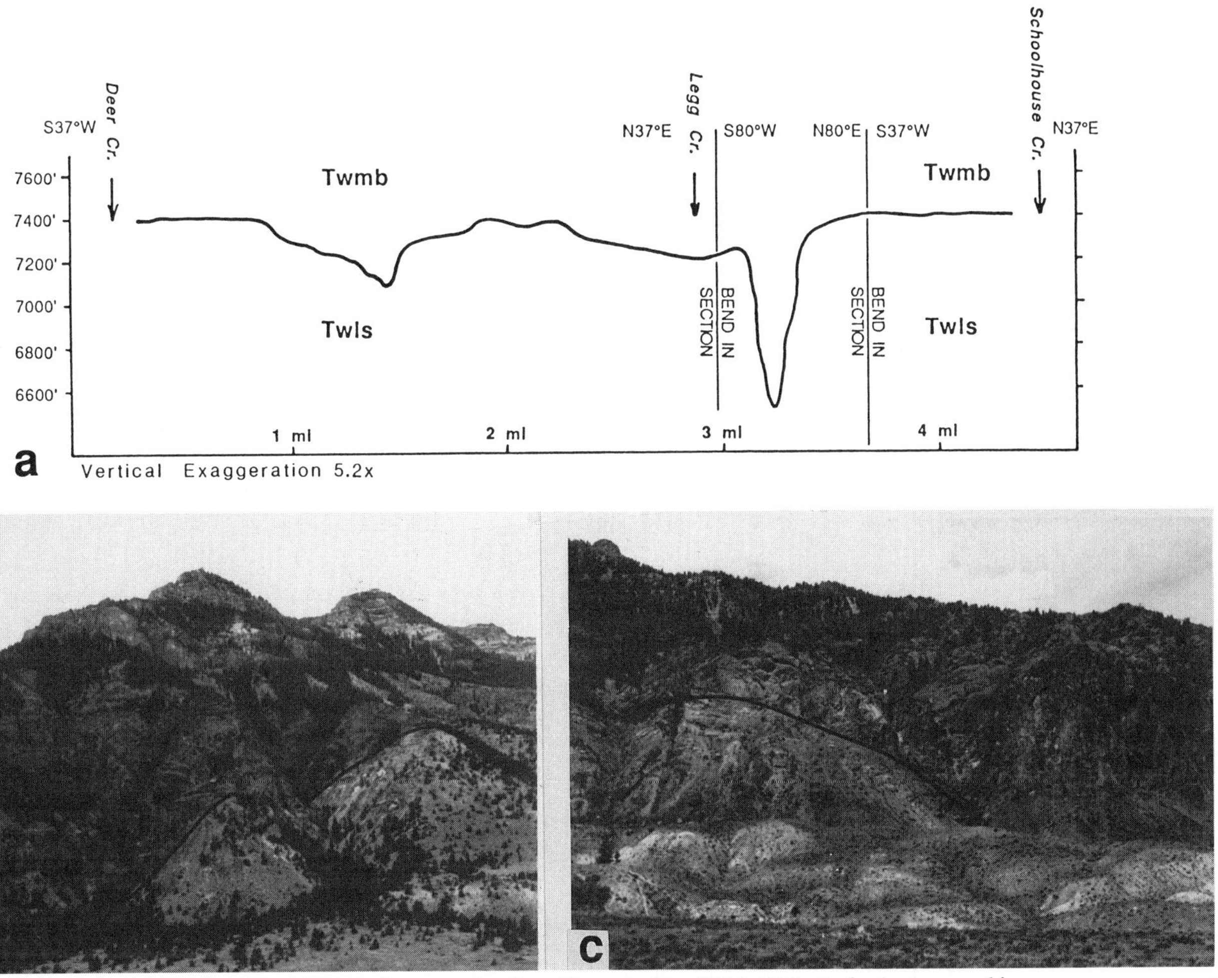

Figure 53. Intra-Wapiti Formation unconformity with as much as 800 ft (250 m) of paleotopographic relief between informally recognized Twls (below) and Twmb (above) members in the upper South Fork Shoshone vally. (a) Profile between Deer Creek and Schoolhouse Creek (see Fig. 1b for location). (b) View to northwest of paleovalley cut in Twls unit, filled with massive breccias of Twmb unit, E of Yellow Creek in Ishawooa valley at UTM 12TXE0114. Maximum relief on unconformity at this locality is roughly 800 ft (240 m). Map pattern appears on Plate I. (c) Paleovalley wall on southeast side of Ishawooa Mesa at UTM 12TXD167989 shows at least 200 ft (60 m) of relief. Erosionally truncated dike (arrow) postdates deposition of Twls unit, but predates Twmb unit; its age must be close to that of the widespread liquefaction-related deformation of the lower AVS, but it has not yet been radiometrically dated.

in these lower AVS units gave rise to deformation, producing structures with a remarkably broad spectrum of styles and orientations within discontinuous areas with poorly defined borders; these areas are classified here into four types of domains on the basis of their predominant structural style.

At the same time, dacitic dikes invaded the epiclastic and pyroclastic section between Deer and Cabin Creeks in the upper South Fork Shoshone River Valley (Plate VII.A), and were deformed during liquefaction prior to becoming rigid. These deformed dacitic rocks were probably precursors to the intrusion of several small quartz diorite plutons in this area, such as the Deer Creek laccolith (Rouse, 1933, 1935). These early dikes may eventually furnish precise constraints on the timing of deformation, but radiometric age analyses are yet to be carried out here.

Relation between liquefaction-related deformation and Heart Mountain faulting

Hauge (1982, 1985) proposed slow gravitational spreading as a solution to the longstanding enigma of the great displacements of Paleozoic carbonate and Eocene volcanic formations by Heart Mountain faulting. However, numerous lines of evidence seem to refute this hypothesis (Pierce, 1982, 1987; Pierce and Nelson, 1986), and the preponderance of evidence favors the traditional interpretation in which short-lived widespread sliding gave rise to a tectonically denuded surface that was buried by Absaroka volcaniclastic debris shortly after faulting (Pierce, 1941, 1957, 1963a, b, 1973, 1982, 1985, 1987; Pierce and Nelson, 1986).

The precise timing of this brief movement on the Heart Mountain fault is of considerable interest in this discussion, because it may correspond very closely to the timing of liquefaction-related deformation in the lowest units of the AVS in the study area. Is this mere coincidence? No direct spatial continuity has yet been demonstrated between these liquefaction-related structures and the Heart Mountain fault. Moreover, even a casual review of the contrasting mechanics and styles of structural development of each would seem at first to preclude any genetic relation between these deformational events. These points notwithstanding, a more careful analysis of timing is warranted by the fact that the bulk of the available evidence suggests that the unusually intense semi-regional liquefaction event identified in this study and the problematic Heart Mountain faulting both occurred in early middle Eocene time. The following argument is based on stratigraphic relations and proposed correlations—regrettably but unavoidably complicated by historical changes in the local stratigraphic nomenclature.

Pierce (1963b) proposed the name Cathedral Cliffs Formation as synonymous with Hague's (1899) and Hague and others' (1896, 1899) informal usage of Early Acid Breccia, and he assigned to it all Absaroka volcanic rocks that pre-date Heart Mountain faulting. Floral evidence suggests that this unit is of early middle Eocene (Bridgerian) age (Pierce, 1963b). Further, Pierce (1963b) recognized the unconformity that separates these light-colored, dominantly pyroclastic rocks from the overlying dark reddish brown, coarse, massive breccias of middle Eocene age (Early Basic Breccia of Hague, 1899, and Hague and others, 1896, 1899), and placed Heart Mountain faulting in the time interval that this unconformity represents.

Hague (1899) mapped no Early Acid Breccia in the South Fork Shoshone valley, and relegated all of the lower Absaroka volcanic-volcaniclastic rocks there to the Early Basic Breccia. In current stratigraphic nomenclature, these rocks are known as Wapiti Formation (Smedes and Prostka, 1972; Pierce and Nelson, 1969). The present study recognizes a major unconformity there within the Wapiti Formation with $\geqslant$800 ft (250 m) of paleotopographic relief carved into the underlying Twls unit (Fig. 53). There are profound contrasts of color and rock type in the Wapiti Formation across this unconformity, between the Twls unit below and the Twmb unit above, just as there are at the unconformity truncating the Cathedral Cliffs Formation. Moreover, on the basis of the marked similarity of only the upper (Twmb) unit to Hague's (1899) description of the Early Basic Breccia (see map unit descriptions, Plate V), and on the tuffaceous character of the lower (Twls) unit, it is proposed that the lowest Absaroka rocks here (the dominantly epiclastic Twls unit) may represent the reworked, medial to distal facies equivalent of the more proximal Cathedral Cliffs Formation pyroclastic rocks. The probable stratigraphic equivalence of these rocks was noted by Wilson (1963, p. 14), and is further supported by the probable Bridgerian ages of both Cathedral Cliffs (Pierce, 1963b) and lower Wapiti/Aycross rocks (Eaton, 1982; Torres, 1985).

If this correlation is valid, it follows that the prelithification deformation of the lower Wapiti Formation and correlative Aycross Formation in the study area occurred simultaneously with Heart Mountain faulting. The mechanism that initiated and sustained the very large, apparently rapid gravitational displacements on the nearly horizontal Heart Mountain fault (Pierce, 1941, 1957, 1963a, b, 1973, 1982, 1985, 1987; Pierce and Nelson, 1986) remains problematic (Piece, 1973, 1987), but a high-energy external stimulus seems necessary to overcome the high initial shear strength and frictional resistance to sliding of the Paleozoic carbonate rocks that make up the major detached masses.

In this dependence on an external trigger, Heart Mountain faulting and the liquefaction-related deformation processes in the east-central Absaroka Range are fundamentally similar, despite their many differences. Perhaps these two deformational events are coeval—but largely spatially and stylistically distinct—first-order responses to a common powerful stimulus, most likely transmitted as seismic energy. The many contrasts between the rigid-body detachment faulting in the northeastern Absaroka Range and the broad range of deformational styles in the study area probably reflect the inherent mechanical differences between the Paleozoic rock units of the former and the largely unconsolidated epiclastic sediments of the latter. Particularly significant would be the seismic propagation properties of each of these media. Additional factors that may have contributed further to the dissimilarity of these responses include the distance of each deformed area from the focus of the triggering seismic energy, and the effects of underlying structure and stratigraphy on the reflection, refraction, and dissipation of that energy.

Middle to late Eocene

Following the erosion of the widely deformed lowest AVS units, the remaining units of the Wapiti and Aycross Formations were deposited, including the Twj, Twmb, Twu, Twuf, and other undivided Tw and Ta rocks. To the north, the coarse, massive breccias of the southward-tapering Twmb unit overwhelmed the significant topography in the area of the present North and South Forks of the Shoshone River (Plate IV). At the same time, equivalent strata within the Aycross Formation may have buried less pronounced, locally developed relief in the Greybull River valley. These proximal to medial breccias gave way toward the close of Wapiti/Aycross time to increasingly more widespread coverage by lavas (Twuf unit). These lavas, interpreted as fissure flows on the basis of their great lateral continuity, pass upward into the Trout Peak Trachyandesite (Ttp), which is of nearly identical composition. All of these lavas appear to have originated at the Sunlight vent complex (Smedes and Prostka, 1972; Sundell and Eaton, 1982).

The youngest AVS units preserved in the area are the middle to late Eocene Blue Point beds (Tbp), the Tepee Trail Formation (Tt), and the Wiggins Formation (Twi). The derivation of these upper AVS units from vent areas to the south and west of the study area (Fig. 1a; Barnes, 1985) documents the southward

Figure 54. Roof contact of felsic stock including Wiggins Formation volcaniclastic strata approximately 1 mi WSW of Frances Peak at UTM 12TXD3167. Relatively more resistant dikes (dark strips) transect both the pluton and the superjacent country rock. View to southwest from summit of Frances Peak.

younging of Absaroka vents along their NNW–SSE trend (Smedes and Prostka, 1972; Love and others, 1976).

Late Eocene or later

Further deformational and intrusive events affected the area following the deposition of the youngest AVS strata here preserved. The radial pattern of a set of porphyritic andesitic dikes south of Greybull River indicates a source of magma at the intermediate to felsic intrusions near Francs Peak, Jojo Mountain, and the Kirwin mineralized area (Fig. 54). Late normal faulting may have been induced by magmatic movements accompanying these intra- or post-Wiggins intrusions. A branching normal fault trends E-W just north of Francs Peak and Jojo Mountain, and cuts at least some of these radial dikes. This fault juxtaposes downthrown vent-facies pyroclastic rocks of the upper Wiggins Formation to the south against the older Aycross, Blue Point, and Tepee Trail rocks to the north (Fig. 55). Displacement is distributed across overlapping fault strands, with an estimated minimum 2400 ft (730 m) of stratigraphic separation.

The minimum age of the steeply dipping N-S– and E-W–trending faults that cut the Wapiti Formation in the Ishawooa area (Plate I) is unconstrained, and it has not been determined whether they are genetically related to the faults in the southern map area. The volcaniclastic breccias and conglomerates in the chaotically dismembered Deer Creek–Cabin Creek domain (Plate VII.A) are cut by a set of closely spaced normal and reverse faults, most of which dip steeply to the southwest, although some dip to the southeast. All of these faults cut through, rather than around, the deposit's cobble-size clasts (Fig. 56), indicating post-lithification movement. Although the offsets on individual faults are small (⩽0.25 m), the cumulative vertical displacement across this zone may be large.

Only one good example of macroscopic folds is present in AVS units of the study area, and its cause and minimum age remain nearly as enigmatic as its singularity. In the Phelps

Figure 55. Normal fault near Gold Reef Tunnel (abandoned mine adit) juxtaposes downdropped Wiggins Formation vent facies breccias against more uniformly stratified Tepee Trail Formation. View to northeast from north face of Francs Peak toward ~UTM 12TXD343710. (Temporary white ground cover at left placed during reclamation phase following recent petroleum exploration drilling.)

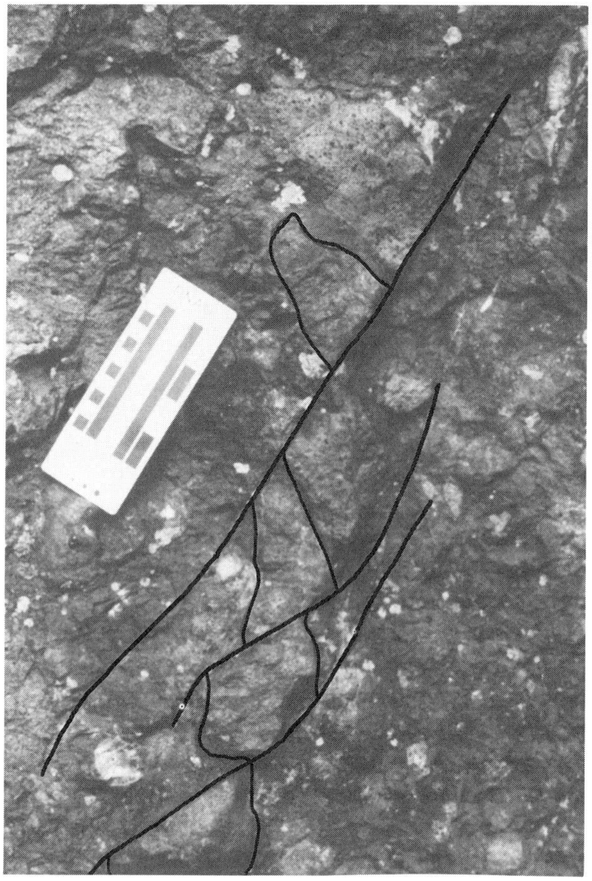

Figure 56. View to northwest of two clasts (one is outlined in ink) offset by several strands of a southwest-dipping reverse fault (heavier inked lines) in Twls breccias approximately 1 mi (1.6 km) north of Cabin Creek at UTM 12TXD088881.

Mountain–Irish Rock area south of the Greybull River, rocks as young as the Tepee Trail and Wiggins Formations are gently folded about slightly northwest-plunging axes in a large anticline-syncline pair (Fig. 57; Plate III). The trend of this flexure is subparallel to that of the Laramide-age Four Bear–Willow Draw anticline nearby, but its late Eocene or younger activity seems to rule out a solely compressional origin. One plausible explanation would be that a Laramide fold, buried beneath the apron of AVS units, was intruded and slightly reactivated by a late Eocene magma body, in much the same way as the dacite intrusion shown in Figure 3 intruded the Four Bear–Willow Draw anticline. Alternatively, intrusion alone may have caused the flexure, although this does not explain the similarity of its trend to that of the Four Bear–Willow Draw structure. Whatever its origin, this anticline was the site of unsuccessful petroleum exploration drilling conducted near Reef Peak on Phelps Mountain in 1985.

CONCLUSIONS

A profusion of different structures testify to highly variable, anomalously intense, early middle Eocene deformation, which compounds the primary depositional complexity of the laterally equivalent, largely epiclastic Aycross and lower Wapiti Formations throughout the east-central Absaroka Range. This disruption is attributed here for the first time to the myriad direct and indirect effects of large-scale, post-depositional liquefaction; seismic energy is recognized as the most probable triggering mechanism.

In many disturbed intervals, present-day stratal geometries and textural configurations mirror little of their primary depositional appearance, reflecting more directly the local response to deformation. In recognition of this fact, continued detailed stratigraphic research in the lower AVS should first question whether the local section is deformed or intact, keeping in mind that both process(es) and severity of deformation may vary radically from one exposure to another. Only after the pre-deformational condition has been confidently reconstructed can investigations proceed to interpret meaningfully the details of the depositional sequence, environment, and process.

Summary of stratigraphic development in the AVS

Despite the complexity introduced by liquefaction and related deformation, both general trends in gross textural character and abundance of primary volcanic material are useful in interpreting the first-order pattern of early-AVS stratigraphic development. Stratigraphic relations among the younger AVS units (both formally and informally recognized) in the study area are much more straightforward by comparison. These units yield more readily to interpretation of depositional environment and process, illuminating significant developmental trends.

The volcanic-volcaniclastic apron of the Absaroka Range accumulated as extrusive activity migrated southeastward during middle to late Eocene time. Thus, the base of the Wapiti Formation climbs progressively higher southeastward, interfingering with and overlying the comparatively distal Aycross Formation. Consistent with this pattern, the predominantly trachyandesite flows of Wapiti/Aycross (Bridgerian) age decrease in thickness and number southward from the Sunlight vent district, with the stratigraphically highest flows (Twuf unit) extending the farthest south, and disappearing entirely in the mountains south of the Greybull valley.

In addition to this first-order pattern of proximal-over-distal facies migration, the Wapiti/Aycross stratigraphic section becomes much thinner, and beds contain more extensively reworked material away from the Sunlight vent area. Relatively fine-grained epiclastic textures increase southward in proportion to primary volcanic and coarse-grained epiclastic textures, reflecting the transition from environments dominated by mass movements and sediment-gravity flows to those in which fluvial deposition prevails. For example, the proximal to medial facies breccias of the middle Wapiti Formation (Twmb unit) taper dramatically southward from more than 1,800 ft (550 m) on Wapiti Ridge to zero thickness on the south flank of Carter Mountain 13 mi (21 km) to the south (Plate IV).

Figure 57. A gently northwest-plunging anticline-syncline pair gently folds all the AVS units present in the Phelps Mountain–Irish Rock area. The cause and youngest possible age of this folding are poorly known. Folding may reflect intrusion of late-stage magma bodies into the cores of buried Laramide anticlines. View to northwest along anticlinal axis (syncline is out of view to right) from top of Phelps Mountain at UTM 12TXD3475.

A tentative revision of lower–AVS correlations in the eastern Absaroka Range proposed here invites more study. The lowest part of the Wapiti Formation in the South Fork Shoshone valley is the dominantly epiclastic Twls unit (informal abbreviation), recognized by its generally well-stratified character, despite locally intense liquefaction and related deformation. This basal AVS unit, and strata of the same age in the Aycross Formation farther south, may represent the medial-distal facies equivalent of the vent–proximal facies Cathedral Cliffs Formation, the basal AVS unit in the northeastern Absaroka Range. This proposed equivalence is based on: (1) similar stratigraphic position with respect to both the bottom of the AVS succession and an overlying unconformity with paleotopographic relief buried by massive volcaniclastic breccias; (2) faunal biostratigraphic evidence, which suggests an early middle Eocene (early Bridgerian) age for both units; and (3) the silicic to intermediate bulk composition and correspondingly light appearance common to both units. If shown to be correct, this stratigraphic reinterpretation bears significantly on the temporal and genetic relations between the liquefaction-related deformation in the study area and Heart Mountain deformation farther north.

Eocene magmatism culminated during Tepee Trail–Wiggins (Bridgerian–Uintan) time with activity along the southern tip of the western chain of Absaroka intrusive/extrusive centers. Vent complexes of this latest stage lie to the south and west of the study area, the closest being the Kirwin mineralized district. Late radial dikes converge on the small intra- or post-Wiggins Formation plutons near Jojo Mountain and Francs Peak, which intruded the volcanic carapace during this final episode.

Summary of early middle Eocene structural development

An original classification scheme describes the exceptionally wide variety of structural styles of early middle Eocene deformation manifest in the Aycross and lower Wapiti Formations. This scheme includes four types of local domains characterized by: tight to isoclinal, inclined to overturned folds (Type 1); layer-parallel shearing structures, including well-developed sheath folds (Type 2); gently dipping truncation surfaces of normal, reverse, and undetermined slip (Type 3); and chaotic dismemberment (Type 4). One of the greatest problems encountered in employing this descriptive regime is inherent in the actual nature of the rock masses it attempts to classify; too commonly, the boundaries between adjacent "domains" of different structural style, or between deformed and essentially undeformed intervals, are arbitrary or simply unmappable because of their complexity. If the resolution of precise domain boundaries is not of primary importance, this scheme, or modifications of it, might be fruitfully employed in other volcanic terranes upon the recognition of similar deformational complexity.

Some of the deformation under investigation here has been recognized but poorly understood for several decades. Hay (1954, 1956) was the first to articulate the significance of fluid behavior in achieving the observed structural relations, but he was unable to offer a satisfying mechanical explanation at that time. More recently, recognition of allochthonous masses on the margins of the Absaroka Range has inspired speculation that this deformation may also have involved large-scale transport of detached rocks. Several lines of reasoning argue strongly against these hypotheses, which include the noncatastrophic "Absaroka Detachment Complex" model proposed by Hauge (1986), the regional gravity-sliding event inferred by Barnes (1985), and the middle Eocene detachment faulting event tentatively suggested by Bown (1982b).

In nearly all occurrences, deformation in domains of Types 1 through 3 is attributed to local gravitational or density instabilities generated or enhanced by liquefaction in one or more nearby Type 4 domains. Most Type 1, 2, or 3 structures constitute evidence of lateral transport, but in no instance are the involved rocks correctly termed allochthonous. In addition, the kinematic patterns inferred from local domains vary nonsystematically across the east-central Absaroka Range. Furthermore, despite the deformation's rootless character, no throughgoing décollement surface or system has yet been discovered, despite its representation on at least one recently published geologic map (Love and Christiansen, 1985). Instead, this early middle Eocene deformation is perhaps best viewed in terms of a set of closely linked in situ disturbances provoked by excess pore fluid pressures.

Whether all deformation within this restricted time interval was initiated in a single liquefaction event, or whether it occurred as a chain of similar events, is not possible to establish firmly from the evidence at hand. However, because earthquakes represent the most viable source of triggering energy for liquefaction across such a large area, it seems intuitively likely that critical pore fluid pressures may have developed repeatedly, as has been documented in seismically active regions in historic times.

The possible stratigraphic equivalence of the deformed interval with the Cathedral Cliffs Formation suggests a previously unobserved relation between this liquefaction-related deformation and Heart Mountain faulting. If confirmed, this equivalence would support the hypothesis proposed here that these two unusual deformational phenomena share a brief time interval and, despite their dramatically different structural styles, may constitute parallel first-order responses to a common causal mechanism. While the mechanical enigma of the Heart Mountain problem has never been adequately resolved, numerous classical interpretations favor a violent high-energy release as the initial motivating impulse. This energy release, or one of similar magnitude from the same source, in the form of earthquake oscillations, may have initiated the liquefaction-induced deformation in the Aycross and lower Wapiti Formations as well. On the other hand, the possibility remains that, whatever triggered Heart Mountain faulting, the actual movement of blocks along the detachment may somehow have caused the deformation described in the area of this study. Nonetheless, it is essential to recognize that, if a genetic link does exist between Heart Mountain faulting and Aycross/Wapiti deformation, it need not be a cause-and-effect relation.

If the proposed correlation is refuted, and the Aycross/Wapiti deformation is in fact younger than Heart Mountain faulting, it may not be necessary to assign these releases of triggering energy to a common source. An alternative explanation would allow the seismic energy source to migrate southward throughout Eocene time, keeping pace with the migration of active magmatism. Hauge (1986), on the basis of his own interpretations of the timing of several deformational events in the Absaroka Range (including Heart Mountain faulting, the Castle Rocks Chaos [Sundell, 1984], the Enos Creek–Owl Creek debris-avalanche [Bown, 1982a, b], and others), asserted that deformation within the AVS progressed southeastward throughout Eocene time. This trend may simply reflect a tendency for freshly deposited units in volcanic terranes to respond to earthquakes caused by local magma movements at depth and eruptions at successively higher stratigraphic levels.

The intrusion of Eocene magma, perhaps into the core of a buried Laramide anticline, may be responsible for the otherwise enigmatic late Eocene(?) gentle folding of AVS strata on the southwestern margin of the study area, in the same way that the dacite body of Figure 3a (unexposed, known only from drilling) locally enhances the closure on the oil-producing Four Bear–Willow Draw anticline in the Greybull valley. This age corresponds closely to the arrival of plutons at this structural level in this part of the Absaroka Range. Other middle to late Eocene (or younger) deformation has introduced significant, but relatively commonplace and geometrically simple, modifications to the architecture of the AVS. These, too, are largely attributable to the effects of Absaroka intrusive and extrusive activity in its waning stages.

In terms of its intensity, the remarkably large area involved, and the diversity of structural styles and deformational processes, the liquefaction-related disruption of early Wapiti/Aycross age is unlike any other event known in the region. Nonetheless, the conditions required for this deformation are not uncommon in rapidly building volcanic terranes of intermediate composition. Additional research on such areas to seek other examples of large-scale, prelithification deformation induced by seismically generated fluid overpressures would be valuable to test the model set forth here.

APPENDIX 1. NOTES ON USE OF UNIVERSAL TRANSVERSE MERCATOR MAP LOCATION SYSTEM

In the interests of precision and efficiency, this report references observation and specimen collection localities in the Universal Transverse Mercator (UTM) grid system. As its name implies, this location system is applicable worldwide, and as a part of the metric system, is already in widespread use around the globe. The UTM grid is typically represented as numbered blue tick marks at the margins of most USGS maps, but many recently published maps provide the full line grid.

Why use the UTM system?

Because daily field work frequently involves the task of precisely locating interesting localities, a convenient, accurate grid system is essential. The well-known systems of geographic (latitude-longitude) coordinates and the Public Land Survey (PLS, or "township-range") grid are both impractical in this regard. On the fine scale required to locate field observations, geographic coordinates are cumbersome for two reasons: (1) the degrees-minutes-seconds of arc hierarchy is constructed in multiples of 60, rather than the more conventional decimal system, and (2) these units of arc are not readily converted to map distances. On the other hand, the widely used PLS grid seems an attractive alternative for the accurate description of tracts on a variety of convenient scales. In practice, however, the PLS system fails in that surveying errors produced a nonuniform grid and, more importantly, that this grid remains unsurveyed in regions encompassing large areas of difficult, undeveloped terrain (as in the Absaroka Range). The utility of this system is further diminished by its application only within parts of the United States. The UTM grid system overcomes all of these problems, and is especially well suited to establishing and referencing locations $\leq$km^2 in size, as is often necessary in geologic field work.

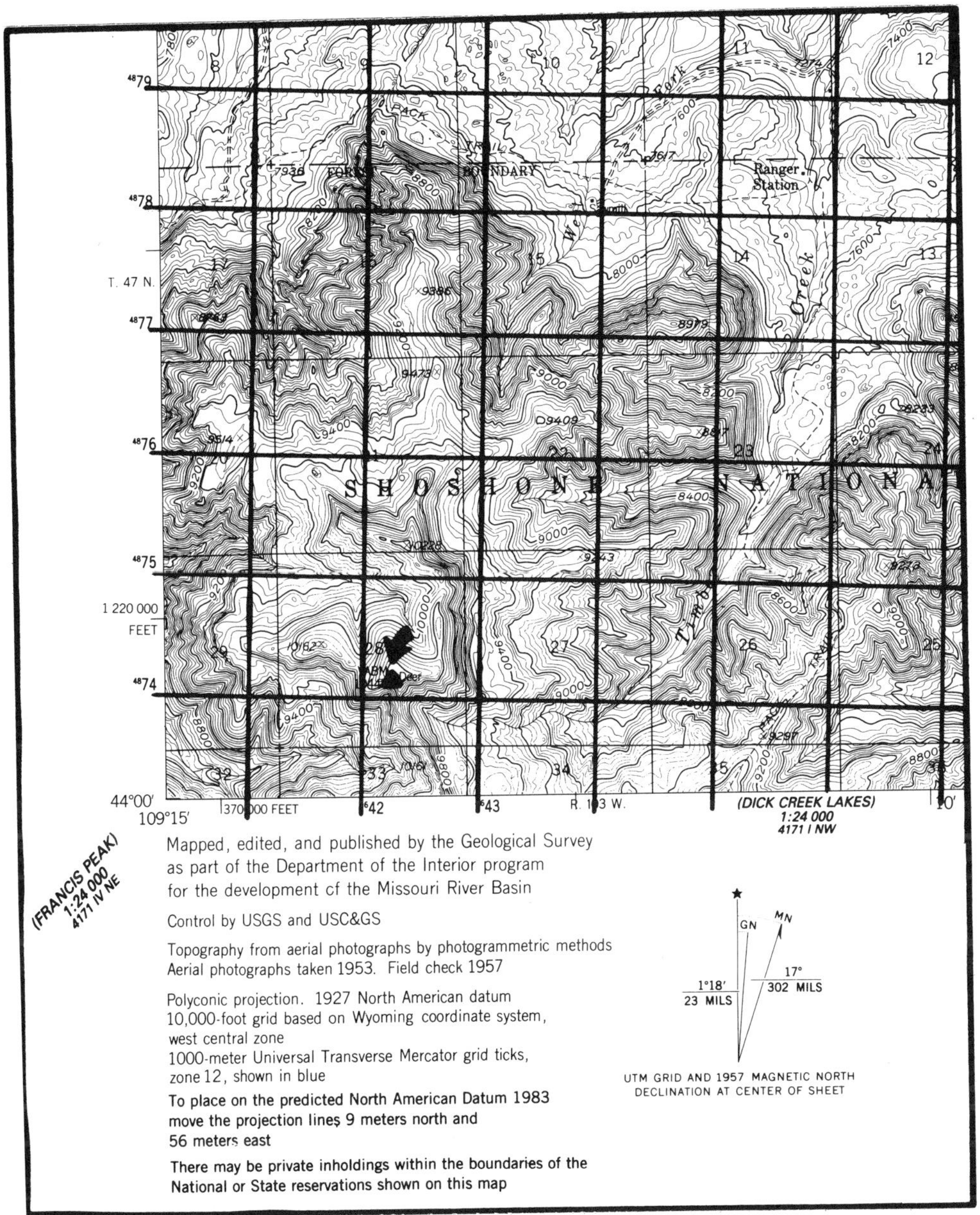

Figure A1. Example of location reference using the Universal Transverse Mercator (UTM) grid system. The summit of Deer Mountain (arrow) lies within 1,000 × 1,000 m (1 km^2) UTM gridsquare 12TXD4274 (specified by 4 digits), and within the smaller 100 × 100 m gridsquare 12TXD422741 (specified by 6 digits). Topographic base is from southwest quarter of USGS Sunshine Reservoir Quadrangle, scale 1:62,500, modified by constructing the full 1,000-m UTM grid and emphasizing the triangle at the summit of Deer Mountain. Timber Creek drainage is to right, Francs Fork drainage to left.

Organization and general application

The following paragraph, excerpted from Merrill (1986), summarizes the organization of the UTM system.

> In the UTM grid system, the Earth is divided into sixty 6° grid zones, numbered consecutively from west to east, beginning at the 180° meridian. The grid zone extending from 174°W to 180° longitude is grid zone 1, and so on. Grid zones extend only from 80°S to 84°N, the polar areas being covered by separate zones. Within each grid zone, a decimal, rectangular grid is fitted whose west-east base line is the equator and whose south-north base line is the central meridian of the grid zone. All south-north grid lines in grid zone 1 are parallel to the 177°W. meridian, and so on. Within each zone, lettered grid subzones (I and 0 are always omitted to avoid confusing them with numbers) extend the full 6° from west to east and 8° (with one exception) from south to north. Each subzone is divided into grid squares that are 100,000 m on a side. The entire global UTM scheme of zones, subzones, and grid squares is illustrated in the U.S. Department of the Army TM-5-241-1 manual (1983) . . . All values in the UTM grid system are read first from west to east and then from south to north. (The military mnemonic of "read RIGHT, UP" is useful.) All designations within the system are for the southwest corner of the selected grid square of whatever size.

The reader is directed to this article (Merrill, 1986) and its list of references for further information on the system's construction and use.

Application to this study

All areas referenced in this study are confined to the 100 × 100-km gridsquares 12TXD and 12TXE, as determined by: (1) reading the latitude-longitude and UTM grid ticks from the USGS topographic quadrangles of the study area, and (2) locating the appropriate gridsquare designation from tables presented in Merrill (1986). It then becomes a simple process to locate a smaller gridsquare within one of these larger areas by adding a numerical suffix. Such a suffix consists of an even number of digits, the first half of which constrains the locality along the W-E axis, and the second half constraining it along the S-N axis. Each digit added to both the W-E and S-N parts of this numerical suffix constrains the locality by an additional order of magnitude in each direction, which narrows down the area to 1/100 of that specified without the added pair of digits. Figure AI.1 illustrates how four digits locates a 1,000 × 1,000-m (1 km^2) gridsquare, while six digits locates a more specific 100 × 100-m gridsquare within it. Similarly, a 10 × 10-m gridsquare may be specified by a suffix of eight digits, and a 1 × 1-m gridsquare by ten digits, although this degree of precision is rarely necessary, and may not be valid, as for example, when areas of this size are beyond map resolution due to small map scale, large contour interval, or lack of topographic detail. Although gridsquares of any size are referenced by the coordinates at their southwest corner, UTM coordinates here are occasionally used to refer only to an observation point within the grid, rather than the entire corresponding gridsquare. All localities described in this volume can be pinpointed using the UTM grid on the appropriate USGS 15-min (1:62,500-scale) and 7½-min (1:24,000-scale) (quadrangles listed in the Introduction, or on the larger 30 × 60-min (1:100,000-scale) Carter Mountain and Ramshorn Quadrangles.

APPENDIX II.
LISTING OF *BASIC* LANGUAGE PROGRAM (INSLOPE modß) USED FOR INFINITE-SLOPE MODEL

```
10 REM ===============================================================
20 REM                              INFINITE SLOPE ANALYSIS
30 REM ===============================================================
40 DEF FNround(x)=INT(x*100+.5)/100
50 REM MODEL ASSUMES: THICKNESS = 100 FT., HT OF GWT = 1/2 * THICKNESS,
51 REM                MOIST DENSITY = 115 PCF, SATURATED DENSITY = 125 PCF,
52 REM                ANGLE OF INTERNAL FRICTION (P) = 25 DEG, COHESION = 0,
53 REM                PSEUDO-ROOT COHESION = 0, SURCHARGE = 0,
54 REM                SEISMIC COEFF. = 0, COEFF. OF FLOW = 8.33 E-9,
55 REM                BOND COEFF. OF MOBILIZED SHEAR STRENGTH = 1/3 * P
56 PRINT
60 REM OPEN OUTPUT FILE FOR STORAGE
70 OPEN "DATAOUT" FOR OUTPUT AS #1
80 REM
100 PRINT "BETA  PHI  H  HW  C  Q   KS   THETO   SAFETY   HC   HR   CREEP RATE"
200 PRINT #1,"BETA  PHI  H  HW  C  Q   KS   THETO   SAFETY   HC   HR   CREEP RATE"
260 FOR slope = 1 TO 20 STEP 2
270 BETA = slope *3.1416/180
280 LET P = 25
290 PHI = P*3.1416/180
300 LET H = 100
310 LET HW = H/2
320 LET GM = 115
330 LET GS = 125
340 LET C = 0
350 LET CR = 0
360 LET Q = 0
370 LET KS = 0
400 LET L = 8.33E-09
410 LET THETO = .3333* P
420 THETO = THETO* 3.1416/180
425 REM  DO CALCULATIONS FOR FACTOR OF SAFETY, ETC.
430 GOSUB 780
435 REM  DO CALCULATIONS FOR CREEP RATE, ETC.
440 GOSUB 950
460 a$="#.###^^^^  #.###^^^^  #.###^^^^  #.###^^^^  #.###^^^^  #.###^^^^  #.###^^^^  #.###^^^^
     #.###^^^^  #.###^^^^  #.###^^^^  #.###^^^^"
470 PRINT USING a$;slope;  P;  H;  HW;  C; Q; KS; THETO; FNround(F); FNround(HWC);
     FNround(HR);   FNround(V)
480 PRINT #1,slope;" "; P;" "; H;" "; HW;" "; C;" "; Q;" "; KS;" "; THETO;" "; FNround(F);" ";
     FNround(HWC);" ";   FNround(HR);" ";   FNround(V)
490 REM
500 NEXT slope
510 REM
540 CLOSE #1
550 END
560 REM===============================================
760 REM SAFETY FACTOR SUBROUTINE. PROGRAM CALCULATES SAFETY
770 REM FACTOR (F) FOR DIFFERENT SOIL AND SLOPE PARAMETERS.
780 C1 = C + CR
785 G1 = (GM*(H-HW)+GS*HW)/H
790 x = C1/(COS(BETA)*COS(BETA)*TAN(PHI))
800 W = (Q+G1*H)*(1-KS*TAN(BETA))+(GS-62.4-GM)*HW
810 Y = (Q+G1*H)*((KS/TAN(BETA))+1)+(GS-GM)*HW
820 Z = (TAN(PHI))/(TAN(BETA))
830 F = ((x+W)/Y)*Z
840 QT = Q+(GM*H)
850 PT = (C1/TAN(PHI))/((COS(BETA))*(COS(BETA)))+QT
860 J = H*G1*KS*TAN(BETA)
870 K = H*G1*KS/TAN(BETA)
880 HWC = (((PT-J)*Z)-(QT+K))/((GS-GM)-(GS-62.4-GM)*Z)
890 RETURN
900 REM===============================================
910 REM CREEP RATE SUBROUTINE. PROGRAM CALCULATES CREEP RATE AT
920 REM SURFACE FOR DIFFERENT SOIL AND SLOPE PARAMETERS. PROGRAM
930 REM REQUIRES INPUT OF COEF. OF FLOW (L) AND BOND COEF. OF SHEAR
940 REM STRENGTH (THETO). EARTHQUAKE COEF. (KS) IS ASSUMED ZERO.
950 THETB = ATN((QT/(PT-62.4*HW))*TAN(BETA))
960 a = ((QT*TAN(BETA))-(PT*TAN(THETO)))/((GM*TAN(BETA))-(GS-62.4)*TAN(THETO))
970 B = (62.4*TAN(THETO))/((GM*TAN(BETA))-(GS-62.4)*TAN(THETO))
980 HR = a+(B*HW)
990 IF HR>=H THEN 1020
1000 V=378000000#*L*(COS(BETA))*(HR/2)*((TAN(THETB))-(TAN(THETO)))
1010 GOTO 1030
1020 V = 378000000#*L*(COS(BETA))*(H/2)*((TAN(THETB))-(TAN(THETO)))
1030 RETURN
```

APPENDIX III. SPECIMENS REFERRED TO IN TEXT.†

Specimen U. W. #	Specimen Field #	Collection Locality (UTM)	Unit	Rock type/ Description	Analytical Procedure(s)*
UW1837-1	86-38	12TXD454737	Tbp	conglomerate matrix/ celadonite mica	XRD, PM
" -2	87-29b	12TXD402931	Twtb	volcanic wacke/ green clay selvage	EDA, PM
" -3	87-26	12TXD351848	Ta	volcanic wacke/ green clay intraclasts	EDA, PM
" -4	86-55b	12TXD422783	Ta	laminated dust tuff/ mm-spaced microfaults	EDA, PM
" -5	86-37a	12TXD456740	Tas	siltst, fine ss/ fossil leaves	---
" -6	87-1	12TXD398782	Ta	silts, mudst/ fossil ferns, reeds	---
" -7	86-22	12TXD477762	Ta	fissile freshwater marl/ fossile gastropods, etc.	---
" -8	86-75	12TXD428784	Ta	welded ash-flow tuff	PM
" -9	86-109a	12TXE195065 (float)	Twls	white tuffaceous sltst, ss	PM
" -10	86-105	12TXE186068	Twmb	from megaclast/ grayish red porphyritic andesite	PM
" -11	86-79	12TXD3659	Twi	rhyolite-andesite vent breccia	---
" -12	86-44	12TXD462745 (float)	Twuf	magnetic porphyritic trachyandesite flow	PM
" -13	87-36	12TXD088880	Td/Twls	dacite dike/ epiclastic ss contact	PM
" -14	86-58	12TXD4775 (float)	Ta	laminated dust tuff/ folds and microfaults	SEM
" -15	87-51	12TXE251028	Twtb	homogeneous epiclastic ss w/ clastic dike filling	PM
" -16	87-40	12TXD103904	Twls	massive epiclastic ss/ clastic dikelet, alteration spots	PM
" -17	87-28	12TXD402931	Twtb	green mudst-sltst-ss inclusion/ clastic dikelets	PM
" -18	86-46	12TXD465747 (float)	Ta	laminated dust tuff/ liquefied ss injections	---
" -19	86-109b	12TXE195065 (float)	Twls	cross-laminated tuffaceous sltst/ dewatering pipes(?)	---
" -20	87-42	12TXE182013 (float)	Tw (in Qls)	epiclastic ss/ hydroplastic slickensides, incipient fabric	PM
" -21	87-31	12TXD402931	Twca pod in Twtb	tuffaceous ss/ anastomosing deformation bands	PM

* Analytical procedures: XRD = X-ray diffraction analysis, EDA = electron microprobe energy dispersive analysis, SEM = scanning electron microscopy, PM = petrographic microscopy.

† All samples are archived in the Geology Museum Repository, Department of Geology and Geophysics, University of Wisconsin-Madison.

REFERENCES CITED

American Geological Institute, 1980, Glossary of geology: American Geological Institute, 749 p.

Aydin, A., and Johnson, A. M., 1983, Analysis of faulting in porous sandstones: Journal of Structural Geology, v. 5, p. 19–31.

Bagnold, R. A., 1954, Experiments on a gravity-free dispersion of large solid spheres in a Newtonian fluid under shear: Proceedings of the Royal Society of London, series A, v. 225, p. 49–63.

Barker, C., 1972, Aquathermal pressuring; Role of temperature in development of abnormal-pressure zones: American Association of Petroleum Geologists Bulletin, v. 56, p. 2068–2071.

Barnes, B. K., 1985, Stratigraphy of Eocene volcaniclastic rocks, Dick Creek Lakes Quadrangle, Wyoming [M.S. thesis]: Ames, Iowa State University, 220 p.

Blackstone, D. L., Jr., 1985, South Fork detachment fault, Park County, Wyoming; Geometry, extent, source: University of Wyoming Contributions to Geology, v. 23, no. 2, p. 47–62.

—— , 1986, Foreland compressional tectonics; Southern Bighorn Basin and adjacent areas, Wyoming: Geological Survey of Wyoming Report of Investigations 34, 32 p.

Bown, T. M., 1980, The Willwood Formation (lower Eocene) of the southern Bighorn Basin, Wyoming, and its mammalian fauna, *in* Gingerich, P. D., ed., Early Cenozoic paleontology and stratigraphy of the Bighorn Basin: University of Michigan Papers on Paleontology, no. 24, p. 127–133.

—— , 1982a, Geology, paleontology, and correlation of Eocene volcaniclastic rocks, southeast Absaroka Range, Hot Springs County, Wyoming: U.S. Geological Survey Professional Paper 1201-A, 75 p.

—— , 1982b, Catastrophic large-scale Late Cenozoic detachment faulting of Eocene volcanic rocks, southeast Absaroka Range, northwest Wyoming: Wyoming Geological Association 33rd Annual Field Conference Guidebook, p. 185–201.

Bucher, W. H., 1933, Volcanic explosions and overthrusts: American Geophysical Union Transactions, 14th Annual Meeting, p. 238–242.

—— , 1935, Remarkable local folding, possibly due to gravity, bearing on the Heart Mountains thrust problem [abs.]: Geological Society of America Proceedings, p. 69.

—— , 1947, Heart Mountain problem: University of Wyoming, Wyoming Geological Association, and Yellowstone-Bighorn Research Association Guidebook to Field Conference in the Bighorn Basin, p. 189–197.

Carmichael, R. S., ed., 1984, CRC handbook of physical properties of rocks, v. 3: Boca Raton, Chemical Rubber Company Press Inc., 340 p.

Carter, N. L., Officer, C. B., Chesner, C. A., and Rose, W. I., 1986, Dynamic deformation of volcanic ejecta from the Toba caldera; Possible relevance to Cretaceous/Tertiary boundary phenomena: Geology, v. 14, p. 380–383.

Castro, G., 1975, Liquefaction and cyclic mobility of saturated sands: American Society of Civil Engineers Journal of the Geotechnical Engineering Division, v. 101, no. GT6, p. 551, 569.

—— , 1987, On the behavior of soils during earthquakes; Liquefaction, *in* Cakmak, A. S., ed., Soil dynamics and liquefaction: Amsterdam, Elsevier, p. 169–204.

Castro, G., and Poulos, S., 1977, Factors affecting liquefaction and cyclic mobility: American Society of Civil Engineers Journal of the Geotechnical Engineering Division, v. 103, no. GT6, p. 501–516.

Chadwick, R. A., 1970, Belts of eruptive centers in the Absaroka-Gallatin volcanic province, Wyoming-Montana: Geological Society of America Bulletin, v. 81, p. 265–274.

Cobbold, P. R., and Quinquis, H., 1980, Development of sheath folds in shear regimes: Journal of Structural Geology, v. 2, p. 119–126.

Coulter, H. W., and Migliaccio, R. R., 1966, Effects of the March 27, 1964, earthquake at Valdez, Alaska: U.S. Geological Survey Professional Paper 542-C, 36 p.

Craig, L. C., Johnson, R. B., Mallory, W. W., McKee, E. D., Roberts, A. E., and Sheldon, R. P., 1972, Mississippian system, *in* Geologic atlas of the Rocky Mountain region: Denver, Colorado, Rocky Mountain Association of Geologists, p. 101–110.

Crozier, M. J., 1986, Landslides; Causes, consequences, and environment: Dover, New Hampshire, Croom Helm, 252 p.

Dake, C. L., 1918, The Hart Mountain overthrust and associated structures in Park County, Wyoming: Journal of Geology, v. 26, p. 45–55.

Davenport, C. A., and Ringrose, P. S., 1987, Deformation of Scottish Quaternary sediment sequences by strong earthquake motions, *in* Jones, M. E., and Preston, R.M.F., eds., Deformation of sediments and sedimentary rocks: Geological Society of London Special Publication 29, p. 299–314.

Decker, P. L., 1987, Structural style of intense gravitational deformation, Absaroka Volcanic Supergroup, Wyoming: Geological Society of America Abstracts with Programs, v. 19, p. 195–196.

—— , 1988, Process and characteristics of liquefaction and cyclic mobility in volcaniclastic sediments; Applications to middle Eocene deformation, Absaroka Range, Wyoming: Geological Society of America Abstracts with Programs, v. 20, p. 412.

Dott, R. H., Jr., and Batten, R. L., 1988, Evolution of the Earth, 4th ed.: New York, McGraw Hill, 704 p.

Eaton, J. G., 1982, Paleontology and correlation of Eocene volcanic rocks in the Carter Mountain area, Park County, southeastern Absaroka Range, Wyoming: University of Wyoming Contributions to Geology, v. 21, p. 153–194.

Erslev, E. A., 1982, The Madison mylonite zone; A major shear zone in the Archean basement of southwestern Montana: Wyoming Geological Association 33rd Annual Field Conference Guidebook, p. 213–221.

—— , 1983, Pre-Beltian geology of the southern Madison Range, southwestern Montana: Montana Bureau of Mines and Geology Memoir 55, 26 p.

Finn, W.D.L., Lee, K. W., and Martin, G. R., 1977, An effective stress model for liquefaction: American Society of Civil Engineers Journal of the Geotechnical Engineering Division, v. 103, no. GT6, p. 517–533.

Fisher, R. V., 1960a, Criteria for recognition of laharic breccias, south Cascade Mountains, Washington: Geological Society of America Bulletin, v. 71, p. 1127–132.

Fisher, R. V., and Schminke, H. U., 1984, Pyroclastic rocks: Berlin, Springer-Verlag, 472 p.

Foose, R. M., Wise, D. U., Fanshawe, J. R., Dutcher, L.A.F., Dutcher, R. R., and French D. E., 1986, Road Log; Red Lodge to Dead Indian Hill via Clarks Fork Canyon; Tectonics and geologic history of the Beartooth uplift, *in* Garrison, P. B. ed., Geology of the Beartooth uplift and adjacent basins: Montana Geological Society and Yellowstone–Bighorn Research Association Joint Field Conference and Symposium, p. 291–306.

Fritz, W. J., and Harrison, S., 1985, Early Tertiary volcaniclastic deposits of the northern Rocky Mountains, *in* Flores, R. M., and Kaplan, S. S., eds., Cenozoic paleogeography of west-central United States; Rocky Mountain Paleogeography Symposium 3: Society of Economic Palentologists and Mineralogists, Rocky Mountain Section, p. 383–402.

Fuller, M. L., 1912, The New Madrid earthquake: U.S. Geological Survey Bulletin 494, 119 p.

Guiraud, M., and Séguret, M., 1987, Soft-sediment microfaulting related to compaction within the fluvio-deltaic infill of the Soria strike-slip basin (northern Spain), *in* Jones, M. E., and Preston, R.M.F., eds., Deformation of sediments and sedimentary rocks: Geological Society of London Special Publication 29, p. 123–136.

Hague, A., 1899, Description of the Absaroka Quadrangle (Crandall and Ishawooa Quadrangles) [Wyoming]: U.S. Geological Survey Geological Survey Geological Atlas, Folio 53.

Hague, A., Weed, W. H., and Iddings, J. P., 1896, Descriptions of the Yellowstone Park Quadrangle (Wyoming): U.S. Geological Survey Atlas, Folio 30.

Hague, A., Iddings, J. P., Weed, W. H., Walcott, C. D., Girty, G. H., Stanton,

T. W., and Knowlton, F. H., 1899, Descriptive geology, petrography, and paleontology; Geology of the Yellowstone National Park, Part 2: U.S. Geological Survey Monograph 32, 893 p.

Hansen, W. R., 1965, Effects of the earthquake of March 27, 1964, at Anchorage, Alaska: U.S. Geological Survey Professional Paper 542-A, 68 p.

Hauge, T. A., 1982, The Heart Mountain detachment fault, northwest Wyoming; Involvement of Absaroka volcanic rock: Wyoming Geological Association 33rd Annual Conference Guidebook, p. 175–179.

—— , 1985, Gravity-spreading origin of the Heart Mountain allochthon, northwestern Wyoming: Geological Society of America Bulletin, v. 96, p. 1440–1456.

—— , 1986, The Absaroka detachment complex, Wyoming; Rootless low-angle normal faulting coeval with Eocene volcanism: Geological Society of America Abstracts with Programs, v. 18, p. 114–115.

Hay, R. L., 1954, Structural relationships of tuff-breccia in Absaroka Range, Wyoming: Geological Society of America Bulletin, v. 65, p. 605–620.

—— , 1956, Pitchfork formation, detrital facies of Early Basic Breccia, Absaroka Range, Wyoming: American Association of Petroleum Geologists Bulletin, v. 40, p. 1863–1898.

Hedge, C. E., Houston, R. S., Tweto, O. L., Peterman, Z. E., Harrison, J. E., and Reid, R. R., 1986, The Precambrian of the Rocky Mountain region: U.S. Geological Survey Professional Paper 1241-D, 17 p.

Hudleston, P. J., 1983, Strain patterns in an ice cap and implications for strain variations in shear zones: Journal of Structural Geology, v. 5, p. 455–463.

Ishihara, K., 1981, Pore water pressure rises during earthquakes: Proceedings of the International Conference of Recent Advances in Geothechnical Earthquake Engineering and Soil Dynamics, St. Louis, v. 3, p. 1201–1204.

Ishihara, K., and Koga, Y., 1981, Case studies of liquefaction in the 1964 Niigata earthquake: Japanese Society of Soil Mechanics and Foundation Engineering Soils and Foundations, v. 21, no. 3, p. 35–52.

Kamada, K., 1980, The Triassic Inai Group in the Karakuwa area, southern Kitakami Mountains, Japan: Journal of Geology, v. 86, p. 713–726 (in Japanese with English abstract and figure captions).

Kawasumi, H., 1968, Introduction, *in* Kawasumi, H., ed., General report on the Niigata earthquake of 1964: Tokyo Electrical Engineering College Press, p. 1–6.

Kobayashi, H., 1968, General reports on damage to buildings in disastered [sic] area, *in* Kawasumi, H., ed., General report on the Niigata earthquake of 1964: Tokyo Electrical Engineering College Press, p. 183–211.

Kraus, M. J., 1985, Early Tertiary quartzite conglomerates of the Bighorn Basin and their significance for paleogeographic reconstruction of northwest Wyoming, *in* Flores, R. M., and Kaplan, S. S., eds., Cenozoic paleogeography of the west-central United States; Rocky Mountain Paleogeography Symposium 3: Society of Economic Paleontologists and Mineralogists, p. 71–91.

Krauskopf, K. B., 1971, Preface, *in* The Great Alaska Earthquake of 1964; Geology: Washington, D.C., National Academy of Sciences, p. ix–xiii.

Kuribayashi, E., and Tatsuoka, F., 1975, Brief review of liquefaction during earthquakes in Japan: Japanese Society of Soil Mechanics and Foundation Engineering Soils and Foundations, v. 15, no. 4, p. 81–92.

Laville, E., and Petit, J. P., 1984, Role of synsedimentary strike-slip faults in the formation of Morroccan Triassic basins: Geology, v. 12, p. 424–427.

Lindsey, D. A., 1972, Sedimentary petrology and paleocurrents of the Harebell Formation, Pinyon Conglomerate, and associated coarse clastic deposits, northwestern Wyoming: U.S. Geological Survey Professional Paper 734-B, 68 p.

Loper, D. E., and McCartney, K., 1988, Catastrophic volcanism as a cause of iridium and shocked quartz found at the K/T boundary [abs.]: EOS Transactions of the American Geophysical Union, v. 69, no. 16, p. 302.

Love, J. D., 1934, The geology of the western end of the Owl Creek Mountains, Wyoming: Wyoming Geological Survey Bulletin, v. 24, 25 p.

—— , 1939, Geology along the southern margin of the Absaroka Range, Wyoming: Geological Society of America Special Paper 20, 136 p.

—— , 1956, New geologic formation names in Jackson Hole, Teton County, northwestern Wyoming: American Association of Petroleum Geologists Bulletin, v. 40, p. 1899–1941.

—— , 1973, Harebell Formation (Upper Cretaceous) and Pinyon Conglomerate (uppermost Cretaceous and Paleocene), northwestern Wyoming: U.S. Geological Survey Professional Paper 734-A, 54 p.

Love, J. D., and Christansen, A. C., 1985, Geologic map of Wyoming: U.S. Geological Survey, scale 1:500,000.

Love, J. D., and Reed, J. C., 1971, Creation of the Teton landscape; The geologic story of Grand Teton National Park: Moose, Wyoming, Grand Teton Natural History Association, 120 p.

Love, J. D., Leopold, E. G., and Love, D. W., 1978, Eocene rocks, fossils, and geologic history, Teton Range, northwestern Wyoming: U.S. Geological Survey Professional Paper 932-B, 40 p.

Love, L. L., Kudo, A. M., and Love, D. W., 1976, Dacites of Bunsen Peak, the Birch Hills, and the Washakie Needles, northwestern Wyoming, and their relationship to the Absaroka volcanic field, Wyoming and Montana: Geological Society of America Bulletin, v. 87, p. 1455–1462.

Mallory, W. W., 1972, Pennsylvanian arkose and the Ancestral Rocky Mountains, *in* Geologic atlas of the Rocky Mountain region: Denver, Colorado, Rocky Mountain Association of Geologists, p. 131–132.

Martin, G. R., Finn, W.D.L., and Seed, H. B., 1975, Fundamentals of liquefaction under cyclic loading: American Society of Civil Engineers Journal of the Geotechnical Engineering Division, v. 101, no. GT5, p. 423–438.

McClay, K. R., and Ellis, P. G., 1987, Geometries of extensional fault systems developed in model experiments: Geology, v. 15, p. 341–344.

McGookey, D. P., Haun, J. D., Hale, L. A., Goodell, H. G., McCubbin, D. G., Weimer, R. J., and Wulf, G. R., 1972, Cretaceous system, *in* Geologic atlas of the Rocky Mountain region: Denver, Colorado, Rocky Mountain Association of Geologists, p. 190–228.

McGrew, A. R., 1965, Stratigraphy and mineralogy of the Blue Point Member of the Wiggins Formation, southeast Absaroka Range, Park County, Wyoming [M.S. thesis]: Laramie, University of Wyoming, 74 p.

Melosh, H. J., 1979, Acoustic fluidization; A new geologic process?: Journal of Geophysical Research, v. 84, p. 7513–7520.

—— , 1981, Acoustically activated décollement; Mechanics of the Heart Mountain fault [abs.]: EOS Transactions of the American Geophysical Union, v. 62, p. 1046.

—— , 1987, The mechanics of large rock avalanches, *in* Costa, J. E., and Wieczorek, G. F., eds., Debris flows/avalanches; Process, recognition, and mitigation: Geological Society of America Reviews in Engineering Geology, v. 7, p. 41–50.

Merrill, G. K., 1986, Map location literacy; How well does Johnny Geologist read?: Geological Society of America Bulletin, v. 97, p. 404–409.

Mueller, P. A., Odom, A. L., and Larsen, L. H., 1976, Archean history of the eastern Beartooth Mountains, Montana: Geological Society of America Abstracts with Programs, v. 8, p. 810.

Nagumo, S., 1968, Field investigations of earthquake phenomena, *in* Kawasumi, H., ed., General report on the Niigata earthquake of 1964: Tokyo Electrical Engineering College Press, p. 157–173.

Nelson, W. H., and Pierce, W. G., 1968, Wapiti Formation and Trout Peak Trachyandesite, northwestern Wyoming: U.S. Geological Survey Bulletin 1354-H, 11 p.

Nelson, W. H., Prostka, H. J., and Williams, F. E., 1980, Geology and mineral resources of the North Absaroka Wilderness and vicinity, Park County, Wyoming: U.S. Geological Survey Bulletin 1447, 101 p.

Nuttli, O. W., 1973, The Mississippi Valley earthquakes of 1811 and 1812; Intensities, ground motion, and magnitudes: Seismological Society of America Bulletin, v. 63, p. 227–248.

Oka, F., Sekiguchi, K., and Goto, G., 1981, Method of analysis of earthquake-induced liquefaction and horizontally layered sand deposits: Japanese Society of Soil Mechanics and Foundation Engineering Soils and Foundations, v. 21, no. 3, p. 7–15.

Owen, G., 1987, Deformation processes in unconsolidated sands *in* Jones, M. E., and Preston, R.M.F., eds., Deformation of sediments and sedimentary rocks: Geological Society of London Special Publication 29, p. 11–24.

Parsons, W. H., 1969, Criteria for the recognition of volcanic breccias; Review, *in* Larsen, L. H., ed., Igneous and metamorphic geology: Geological Society of America Memoir 115, p. 263–304.

——, 1974, Volcanic rocks of the Absaroka–Yellowstone region, *in* Voight, B., ed., Rock mechanics; The American Northwest; Third International Congress on Rock Mechanics: University Park, Pennsylvania State University College of Earth and Mineral Sciences, p. 94–101.

——, 1978, Middle Rockies and Yellowstone: Minneapolis, Kendall Hunt Publishing, 233 p.

Penick, J. L., Jr., 1976, The New Madrid earthquakes of 1811–1812: Columbia, University of Missouri Press, 181 p.

Peterson, J. A., 1972, Jurassic system, *in* Geologic atlas of the Rocky Mountain region: Denver, Colorado, Rocky Mountain Association of Geologists, p. 177–189.

——, 1987, Pre-thrusting regional stratigraphy and paleotectonics, northern Rocky Mountains and adjacent Cordilleran shelf, U.S.: Wyoming Geological Association 38th Field Conference Guidebook, p. 17–26.

Petit, J. P., and Laville, E., 1987, Morphology and microstructures of hydroplastic slickensides in sandstones, *in* Jones, M. E., and Preston, R.M.F., eds., Deformation of sediments and sedimentary rocks: Geological Society of London Special Publication 29, p. 107–121.

Pierce, W. G., 1941, Heart Mountain and South Fork thrusts, Park County, Wyoming: American Association of Petroleum Geologists Bulletin, v. 25, p. 2021–2045.

——, 1957, Heart Mountain and South Fork detachment thrusts of Wyoming: American Association of Petroleum Geologists Bulletin, v. 41, p. 591–626.

——, 1963a, Reef Creek detachment fault, northwestern Wyoming: Geological Society of America Bulletin, v. 74, p. 1225–1236.

——, 1963b, Cathedral Cliffs Formation; The Early Acid Breccia unit of northwest Wyoming: Geological Society of America Bulletin, v. 74, p. 9–22.

——, 1970, Geologic map of the Devils Tooth Quadrangle, Park County Wyoming: U.S. Geological Survey Geologic Quadrangle Map GQ–817, scale 1:62,500.

——, 1973, Principal features of the Hearth Mountain fault and the mechanism problem, *in* DeJong, K. A., and Scholten, R., eds., Gravity and tectonics: New York, John Wiley and Sons Interscience, p. 457–472.

——, 1978, Geologic map of the Cody 1° × 2° Quadrangle, Wyoming: U.S. Geological Survey Miscellaneous Investigations Map MF–963, scale 1:250,000.

——, 1982, Relation of volcanic rocks to the Heart Mountain fault: Wyoming Geological Association Guidebook 33rd Annual Conference, p. 181–183.

——, 1985, Map showing present configuration of Heart Mountain fault and related features, Wyoming and Montana: Geological Survey of Wyoming MS–15, scale 1:125,000.

——, 1986, The South Fork detachment fault, Park County, Wyoming; Discussion and reply: University of Wyoming Contributions to Geology, v. 24, no. 1, p. 77–90.

——, 1987, A case for tectonic denudation by the Heart Mountain fault; A response: Geological Society of America Bulletin, v. 99, p. 552–568.

Pierce, W. G., and Andrews, D. A., 1941, Geology and oil and coal resources of the region south of Cody, Park County, Wyoming: U.S. Geological Survey Bulletin 921–B, p. 99–180.

Pierce, W. G., and Nelson, W. H., 1969, Geologic map of the Wapiti Quadrangle, Park County, Wyoming: U.S. Geological Survey Geological Quadrangle Map GQ–778, scale 1:62,500.

——, 1973, Crandall Conglomerate; An unusual stream deposit, and its relation to Heart Mountain faulting: Geological Society of America Bulletin, v. 84, p. 2631–2644.

——, 1986, Some features indicating tectonic denudation by the Heart Mountain fault, *in* Garrison, P. B., ed., Geology of the Beartooth uplift and adjacent basins: Montana Geological Society and Yellowstone–Bighorn Research Association Joint Field Conference and Symposium, p. 155–164.

Pierson, T. C., 1984, A rheologic classification of subaerial sediment-water flows: Geological Society of America Abstracts with Programs, v. 16, p. 623.

Pierson, T. C., and Costa, J. E., 1987, A rheologic classification of subaerial sediment-water flows, *in* Costa, J. E. and Wieczorek, G. F., eds., Debris flows/avalanches; Process, recognition, and mitigation: Geological Society of America Reviews in Engineering Geology, v. 7, p. 1–12.

Rice, A., 1985, Popcorn magma: EOS Transactions of the American Geophysical Union, v. 66, no. 18, p. 392.

——, 1986a, Quench supersaturation and volcanic eruptions [abs.]: The Journal of the Colorado–Wyoming Academy of Science, v. 18, no. 1, p. 37.

——, 1986b, Detonation pressures in explosive volcanism; A source for shocked minerals . . . and whole mantle convection; A source of explosive volcanism at the end of the Cretaceous [abs.]: EOS Transactions of the American Geophysical Union, v. 67, no. 44, p. 1258.

Rohrer, W. L., 1966a, Geology of the Adam Weiss Peak Quadrangle, Hot Springs and Park Counties, Wyoming: U.S. Geological Survey Bulletin 1241–A, p. 39.

——, 1966b, Geologic map of the Kisinger Lakes Quadrangle, Fremont County, Wyoming: U.S. Geological Survey Geologic Quadrangle Map GQ–527, scale 1:24,000.

Rohrer, W. L., and Obradovich, J. D., 1969, Age and stratigraphic relations of the Tepee Trail and Wiggins Formations, northwestern Wyoming: U.S. Geological Survey Professional Paper 650–B, p. B57–B62.

Ross, G. A., Seed, H. B., and Migliaccio, R. R., 1969, Bridge foundations in Alaska earthquake: American Society of Civil Engineers Journal of the Soil Mechanics and Foundations Division, v. 95, no. SM4, p. 1007–1036.

Rouse, J. T., 1933, The structure, inclusions, and alteration of the Deer Creek intrusive, Wyoming: American Journal of Science, v. 26, p. 139–146.

——, 1935, The volcanic rocks of the Valley area, Park County, Wyoming: EOS Transactions of the American Geophysical Union, v. 116, p. 274–284.

——, 1937, Genesis and structural relationships of the Absaroka volcanic rocks, Wyoming: Geological Society of America Bulletin, v. 48, p. 1257–1296.

——, 1940, Structural and volcanic problems in the southern Absaroka Mountains, Wyoming: Geological Society of America Bulletin, v. 51, p. 1413–1428.

Schmitt, J. G., 1987, Origin of Late Cretaceous to early Tertiary quartzite conglomerates in northwestern Wyoming: Wyoming Geological Association 38th Field Conference Guidebook, p. 217–224.

Seed, H. B., 1979, Soil liquefaction and cyclic mobility evaluation for level ground during earthquakes: American Society of Civil Engineers Journal of the Geotechnical Engineering Division, v. 105, no. GT2, p. 201–255.

Seed, H. B., and Idriss, I. M., 1982, Ground motions and soil liquefaction during earthquakes: Berkely, California, Earthquake Engineering Research Institute, 134 p.

Seed, H. B., and Wilson, S. D., 1967, The Turnagain Heights landslide, Anchorage, Alaska: American Society of Civil Engineers Journal of Soil Mechanics and Foundations Division, v. 93, no. SM4, p. 325–353.

Seeland, D., 1978, Sedimentology and stratigraphy of the lower Eocene Wind River Formation, central Wyoming: Wyoming Geological Association 30th Annual Field Conference Guidebook, p. 181–198.

Shive, P. N., and Pruss, E. F., 1977, A paleomagnetic study of basalt flows from the Absaroka Mountains, Wyoming: Journal of Geophysical Research, v. 82, p. 3039–3048.

Sloss, L. L., 1963, Sequences in the cratonic interior of North America: Geological Society of America Bulletin, v. 74, p. 93–114.

Smedes, H. W., 1966, Geology and igneous petrology of the northern Elkhorn Mountains, Jefferson and Broadwater Counties: Montana: U.S. Geological Survey Professional Paper 510, 116 p.

Smedes, H. W., and Prostka, H. J., 1972, Stratigraphic framework of the Absaroka Volcanic Supergroup in the Yellowstone National Park Region: U.S. Geological Survey Professional Paper 729–C, p. C1–C33.

Smith, G. A., 1986, Coarse-grained nonmarine volcaniclastic sediment; Terminology and depositional process: Geological Society of America Bulletin, v. 97, p. 1–10.

Steidtmann, J. R., 1971, Origin of the Pass Peak Formation and equivalent early Eocene strata, central western Wyoming: Geological Society of America Bulletin, v. 82, p. 156–176.

Sundell, K. A., 1984, The Castle Rocks chaos; A gigantic Eocene landslide-debris flow within the southeastern Absaroka Range, Wyoming [Ph.D. thesis]: University of California at Santa Barbara, 383 p.

Takahashi, T., 1978, Mechanical characteristics of debris flows: American Society of Civil Engineers Journal of the Hydraulics Division, v. 104, no. HY8, p. 1153–1169.

Ter-Stepanian, G., 1963, On the long-term stability of slopes: Norwegian Geotechnical Institute Publication 52, p. 1–14.

——, 1974, Depth creep of slopes: Bulletin of the International Association of Engineering Geology, no. 9, p. 97–102.

Tohno, I., and Yasuda, S., 1981, Liquefaction of the ground during the 1978 Miyagiken-Oki earthquake: Japanese Society of Soil Mechanics and Foundation Engineering Soils and Foundations, v. 21, no. 3, p. 18–34.

Torres, V., 1985, Stratigraphy of the Eocene Willwood, Aycross, and Wapiti formations along the North Fork of the Shoshone River, north-central Wyoming: University of Wyoming Contributions to Geology, v. 23, no. 2, p. 83–97.

Torres, V., and Gingerich, P. D., 1983, Summary of Eocene stratigraphy at the base of Jim Mountain, North Fork of the Shoshone River, northwestern Wyoming: Wyoming Geological Association 34th Annual Field Conference Guidebook, p. 205–208.

Updike, R. G., Eagan, F. A., Yoshiharu, M., Idriss, I. M., and Moses, T. L., 1988, A model for earthquake-induced translatory landslides in Quaternary sediments: Geological Society of America Bulletin, v. 100, p. 783–792.

Van Bemmelen, R. W., 1949, Geology of Indonesia: The Hague, The Netherlands, Government Printing Office, 732 p.

Van Houten, F. B., 1944, Stratigraphy of the Willwood and Tatman Formations in northwestern Wyoming: Geological Society of America Bulletin, v. 55, p. 165–510.

Veyera, G. E., and Charlie, W. A., 1987, Liquefaction of shock loaded saturated sand, *in* Cakmak, A. S., ed., Soil dynamics and liquefaction: Amsterdam, Elsevier, p. 205–219.

Wang, Z. Q., 1981, Macroscopic approach to soil liquefaction: Proceedings of the International Conference of Recent Advances in Geotechnical Earthquake Engineering and Soil Dynamics, St. Louis, v. 1, p. 179–185.

Wilson, G., 1982, Introduction to small-scale geologic structures: London, George, Allen, and Unwin, 128 p.

Wilson, W. H., 1963, Correlation of volcanic rock units in the southern Absaroka Mountains, northwest Wyoming: University of Wyoming Contributions to Geology, v. 2, p. 13–20.

——, 1964a, Geologic reconnaissance of the Absaroka Mountains, northwest Wyoming; Part 1, The Wood River–Greybull River area: University of Wyoming Contributions to Geology, v. 3, p. 60–77.

——, 1964b, The Kirwin mineralized area, Park County, Wyoming: Wyoming Geological Survey Preliminary Report 2, 12 p.

——, 1975, Detachment faulting in volcanic rocks, Wood River area, Park County, Wyoming: Wyoming Geological Association 27th Annual Field Conference Guidebook, p. 167–171.

——, 1982, Geology of the Dick Creek Lakes, Dunrud Peak, Francs Peak, Noon Point, and Twin Peaks Quadrangles, Fremont, Hot Springs, and Park Counties, Wyoming: Geological Survey of Wyoming Map Series 10, scale 1:50,000.

Wong, R. R., Seed, G. B., and Chan, C. K., 1975, Cyclic loading liquefaction of gravelly soils: American Society of Civil Engineers Journal of the Geotechnical Engineering Division, v. 101, no. GT6, p. 571–583.

Manuscript Accepted by the Society June 1, 1989

Typeset by WESType Publishing Services, Inc., Boulder, Colorado
Printed in U.S.A. by Malloy Lithographing, Inc., Ann Arbor, Michigan